Cell membranes and ion transport

Integrated themes in biology

Consulting Editor: I. D. J. Phillips, University of Exeter

Hall and Baker: Cell Membranes and Ion Transport

Pitt: Lysosomes and Cell Function

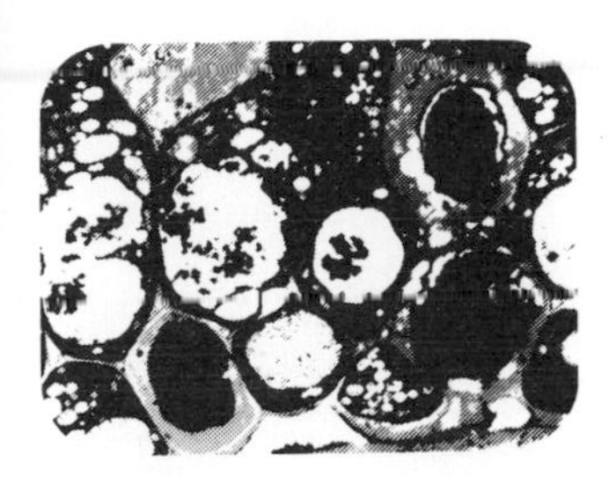

Cell membranes and ion transport

J. L. Hall and D. A. Baker

School of Biological Sciences
University of Sussex

Longman London and New York

Longman Group Limited London

Associated companies, branches and representatives throughout the world

Published in the United States of America by Longman Inc., New York

First published 1977

Library of Congress Cataloging in Publication Data

Hall, John Lloyd.
Cell membranes and ion transport.

(Integrated themes in biology)
Includes index.
1. Biological transport. 2. Cell membranes.
I. Baker, D. A., joint author. II. Title.
QH509.H36 574.8'75 77-3142
ISBN 0-582-44192-7

Printed in Great Britain by Whitstable Litho Ltd., Whitstable, Kent

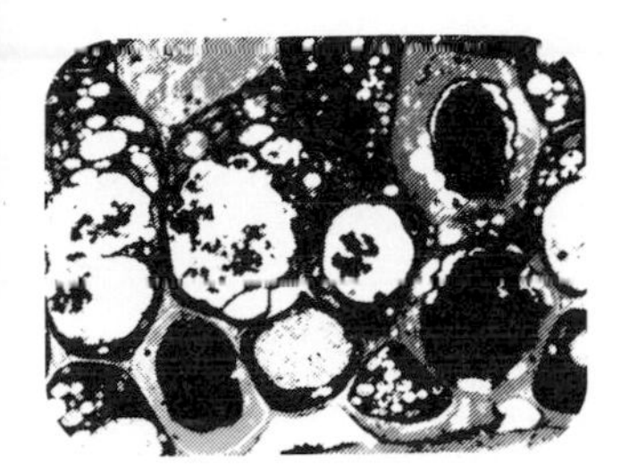

Preface

The study of biological membranes and ion transport is a vast and complicated subject. In this short book, we have attempted to integrate information and concepts emerging from current research in this field. This information, taken from studies of animal, plant and bacterial cells, is presented here at a level suitable for first and second year undergraduate students of biological sciences. The separate chapters on structure and properties of membranes, biophysical aspects of ion transport, passive permeation, and the linkage of ion transport to metabolism, are all closely interrelated. We hope that readers of this book will also interrelate the information and thus perceive the membrane as a functional unit, the complex biology of which is fundamental to the integrity of life itself. To facilitate this understanding, we have presented two case studies, one plant and one animal, which are intended to illustrate many of the principles introduced in the earlier chapters. Some familiarity with related biophysics and biochemistry is necessary before these case studies can be approached, and it is therefore advisable to read the first four chapters before attempting to understand the integrated transport and metabolic processes described in these systems.

The literature citations are not intended to provide a full coverage, although a number of interesting papers are quoted in each chapter. We hope that some of our readers will find the subject of sufficient interest to wish to pursue it further, and so we have suggested some up-to-date books and reviews which cover certain important aspects in greater detail. In writing these chapters, we have been conscious of how much research is still needed to achieve a full understanding of membrane function. Our present knowledge provides, at best, some indication of the problems still to be resolved and perhaps a glimpse of the possibilities which exist beyond the boundaries of current research.

We would like to thank the many authors who have generously supplied data and illustrations. Our thanks are also due to Drs C. D. Field, R.

Jennings, and K. P. Wheeler for valuable discussions of the drafts of the chapters, to Miss N. Browning for typing the manuscript and to Mr. G. W. Trevelyan for help in editing the manuscript. Any faults or omissions that remain are our own. J.L.H. D.A.B. *Sussex*, 1977

Acknowledgements

We are grateful to the following for permission to reproduce copyright material: Academic Press, Inc. and authors, P. P. H. DeBruyn, S. Michelson and R. P. Becker for Figs. 18 and 34 – *Journal of Ultrastructure Research* (1975); American Association for the Advancement of Science for Fig. (p. 329) B. H. J. Hofstee – *Science* Vol. 116 (1952) and Figs. 2 and 3, S. J. Singer and G. L. Nicholson, *Science*, Vol. 175 (1972); The American Society of Plant Physiologists for Fig. 4, T. Kagawa *et al.* – *Plant Physiology* Vol. 51 (1973); American Society of Zoologists for Fig. 6, C. L. Slayman – *American Zoologist* Vol. 10 (1970); Annual Reviews, Inc., for Fig. 1, S. J. Singer and Fig. (p. 337) J. L. Dahl and L. E. Hokin – *Annual Review of Biochemistry* Vol. 43 (1974); The Biochemical Journal and authors, R. Whittam and M. E. Ager for Figs. 1 and 2 – *Biochemical Journal* Vol. 9 (1965) and Fig. 1, R. Whittam and D. M. Blond – *Biochemical Journal* Vol. 92 (1964); Blackwell Scientific Publications Ltd., for Fig. 4, F. A. Smith – *New Phytologist* Vol. 69 (1970) and Fig. 7, G. E. Briggs *et al.* – *Electrolytes and Plant Cells* (1961); British Council Medical Department and author J. A. Lucy for Fig. 1 – *British Medical Bulletin* Vol. 24 No. 2, reproduced by permission; Brookhaven Natl. Laboratories for Fig. 9 (top part) – A. T. Jagendorf and E. Urise – *Brookhaven Symp. Biology* Vol 19 (1967); Cambridge University Press and authors, A. B. Hope and N. A. Walker for Fig. 11.2 – *The Physiology of Giant Algal Cells* (1975) and for Fig. 1. R. N. Robertson – *Protons, Electrons, Phosphorylation and Active Transport* (1968); Elsevier North-Holland Biomedical Press for Fig. 3, J. C. Skou – *Biochim. Biophys. Acta* (1960), Figs. 2 and 6, R. K. Crane – *Biochim. Biophys. Acta* (1965) and Fig. 3.3, J. B. Hanson and D. E. Koeppe – *Ion Transport in Plant Cells and Tissues* (1975) eds. Baker and Hall; Federation of American Societies for Experimental Biology for Figs. 1 and 2, R. K. Crane *et al.* – *Federation Proceedings* 24, (1965); Harper & Row Publishers, Inc., for Fig. 18, J. D. Robertson – *Molecular Organisation and Biological Function* (1967) ed. J. M. Allen; The Journal of Physiology and authors for Fig. 4, R. Whittam, Vol. 140 (1958), Fig. 4, P. C. Caldwell *et al.* Vol. 152 (1960) and Fig. (p. 486) E. M. Wright Vol. 185 (1966) – *Journal of Physiology*; Longman Group Ltd., for Figs. 2.16, 8.9 and 8.18, J. L. Hall, T. J. Flowers and R. M. Roberts – *Plant Cell Structure and Metabolism* (1974); Macmillan Journals Ltd., and authors, S. G. Schultz and R. Zalusky for Fig. 1 – *Nature* Vol. 205 (1965); McGraw-Hill Book Company (UK) Ltd., for Figs. 3.18 and 3.19, D. J. Morre and H. H. Mollenhauer – *Dynamic Aspects of Plant Ultrastructure*, ed. A. W. Robards (1974) © 1974 McGraw-Hill Book Co. (UK) Ltd., reproduced by permission; Oxford University Press for Fig. (p. 1) W. J. Lucas and F. A. Smith – *Journal Experimental Botany* Vol. 24 (1973); Scandinavian Society for Plant Physiology for Fig. 3, R. Collander – *Physiologia Planterum* (1954); Sinauer Associates, Inc. and authors, P. D. Boyer and W. L. Klein for Fig. 6 – *Membrane Molecular Biology*, eds. Fox and Keith (1972); Springer-Verlag, Inc., for Figs. 3 and 8, A. E. Hill – *Journal Membrane Biology* Vol. 12 (1973) and Fig. 12*g*, D. F. H. Wallach – *The Plasma Membrane* (1975); Symposia of the Society for Experimental Biology for Fig. 3 H. R. Kaback and Fig. 1 and 2, J. F. Danielli – *Symposia for Experimental Biology* (1973) and (1954); Wistar Institute of Anatomy and Biology for Fig. 1, J. Danielli and H. Davson – *Journal of Cellular and Comparative Physiology* (1935).

Whilst every effort has been made to trace the owners of copyright, in a few cases this has proved impossible and we take this opportunity to offer our apologies to any authors whose rights may have been unwittingly infringed.

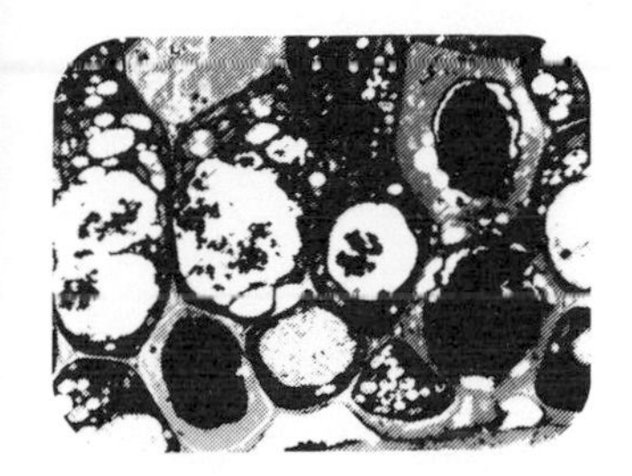

Contents

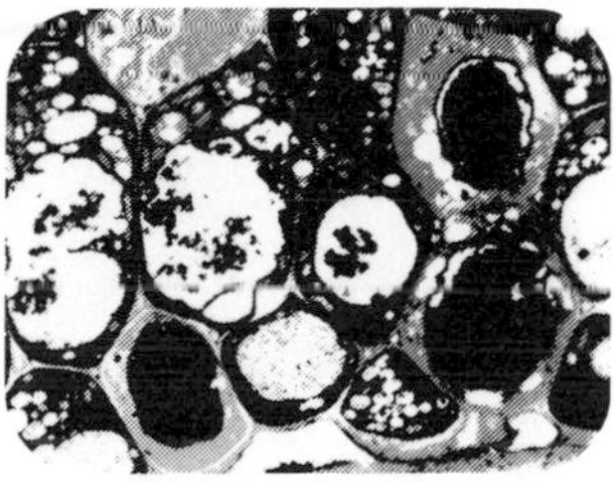

Chapter 1

Structure and properties of cell membranes

All living cells are delimited from their environment by a surface membrane, the plasma membrane or plasmalemma, while eukaryotic plant and animal cells are further compartmentalized intracellularly into various membrane-bound organelles (Fig. 1.1). Perhaps the most fundamental property of these membranes is their ability to act as selective permeability barriers, controlling the amount and nature of the substances that pass across them. The development of such a boundary was essential to the development of primitive cells, which presumably arose by the formation of a membrane around a few macromolecules possessing catalytic properties, thus allowing other smaller molecules and ions to interact. Such a membrane would need to possess a certain selectivity allowing the cell to develop characteristic properties; it would permit the selection and retention of essential molecules that became depleted and the removal of waste products of metabolism. Such transport systems must have appeared quite early in evolution, perhaps before the appearance of metabolic pathways (see Holden, 1968; Pardee and Palmer, 1973). Today, organelles and pleuropneumonia-like organisms which live surrounded by cytoplasm possess membranes and transport systems but lack a complete enzymic machinery, an observation consistent with the above idea. If this is the case, then presumably membranes evolved even earlier. In addition, such a cell would need the ability to respond to changes in the external medium. For example, the cell volume must remain fairly constant, since large changes would affect the concentration of molecules within the cell and so the interactions between them. This could be achieved by regulation of the concentration of some internal osmotic agent to counteract the changes occurring in the environment. Today, many animal cells regulate their osmotic environment by regulating the internal concentration of sodium ions.

The permeability of cell membranes to different substances varies widely. Gases move across membranes quite easily and small molecules pass through more easily than larger molecules with similar chemical

(a)

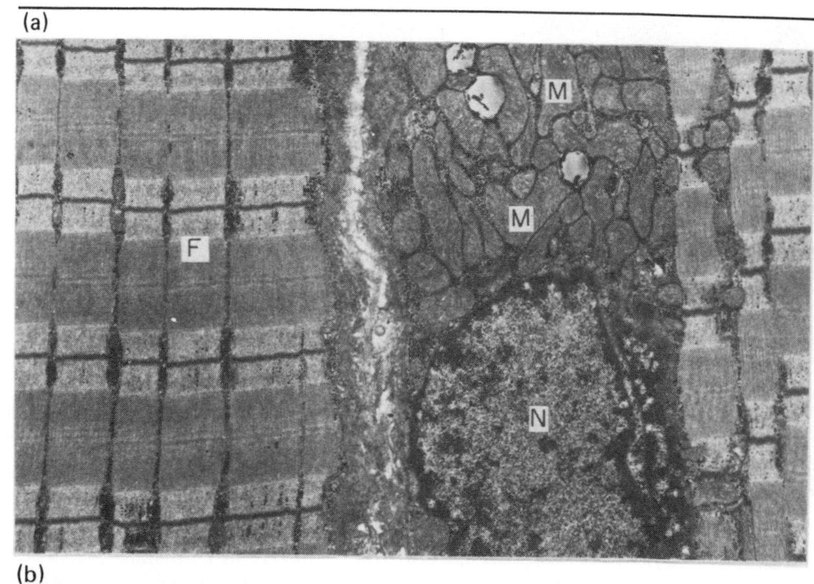

(b)

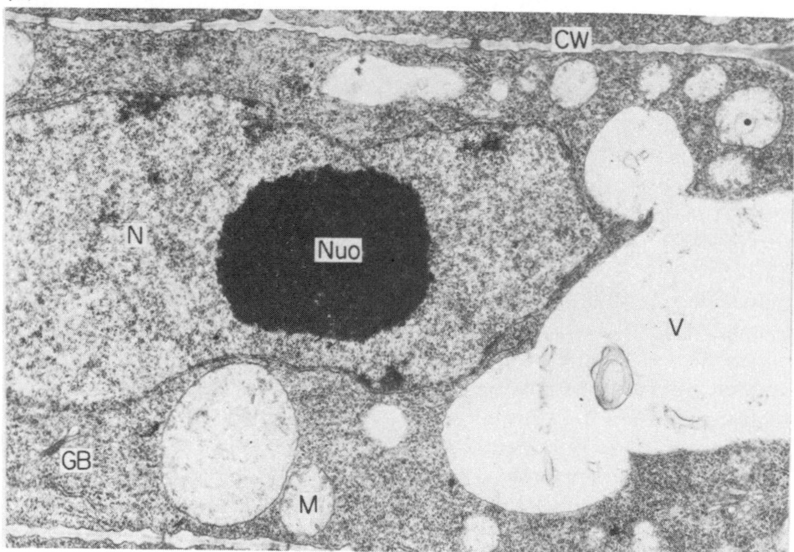

Fig. 1.1 Electron micrographs of animal and plant cells showing a variety of membrane-bound organelles. (*a*) Rat hind limb skeletal muscle, × 10,000. (*b*) Root tip cell of the halophyte *Suaeda maritima*, × 11,500. Note the presence of the cell wall and large vacuoles in the plant cell. N, nucleus; Nuo, nucleolus; M, mitochondrion; F, muscle fibre; V, vacuole; CW, cell wall; GB, Golgi body. ((*a*) was provided by Dr Gillian R. Bullock, CIBA Laboratories, Horsham, UK.)

Table 1.1 Concentrations of potassium, sodium and chloride in various cells and in the external medium.

	Concentration of ions (mol m^{-3})		
	K^+	Na^+	Cl^-
(*a*) Mammalian muscle			
Interstitial fluid	4	145	120
Intracellular fluid	155	12	4
(*b*) Squid axon			
External medium	10	468	540
Axon	400	50	50
(*c*) *Acetabularia mediterranea*			
External solution	10	470	550
Cytoplasm	400	57	480
Vacuole	355	65	480
(*d*) Washed carrot tissue			
Bathing medium	5	5	20
Cell sap	85	23	19

((*a*) From Nystrom, 1973; (*b*) from Kotyk and Janáček, 1970; (*c*) from Saddler, 1970; (*d*) from Cram, 1975.)

properties. Usually the greater the lipid solubility of a substance, the greater its permeability through the membrane, while cells are usually less permeable to electrolytes than non-electrolytes (see p. 60). Substances may diffuse through membranes passively or may require the expenditure of energy to achieve this movement. The selectivity of membranes prevents the random diffusion of substances and allows the cell to determine and control its internal environment while providing a number of different micro-environments within the subcellular organelles. Membranes allow the cell or cell organelle to be isolated from its surroundings and so allow the highly ordered biochemical processes of life to go on relatively unaffected by changes in the environment. At the same time, membranes allow a constant exchange of substances with the outside medium, including the uptake of oxygen and essential nutrients and the exclusion of unwanted and harmful materials. The control of ion movement is clearly of fundamental importance in this homeostatic process since it will directly affect such factors as cellular pH and various osmotic and metabolic functions. Animal cells, in particular, show remarkably consistent electrolyte contents, and both plant and animal cells maintain concentrations of individual ions which are very different from those in the extracellular environment (Table 1.1). The most important membrane in this respect is the plasma membrane, which is the primary barrier to the movement of ions into and out of cells. In vacuolated plant cells, the tonoplast or vacuolar membrane provides a second important barrier, since the vacuole is a major site of ion accumulation. In addition, the ion content of cell organelles may show marked differences from that of other parts of the cell and the environment. Thus, when we consider the total ionic relations of a cell or tissue, it must be remembered that more than one membrane barrier is involved, the whole providing a complex system of selectively permeable barriers. However, although we will be largely concerned here with the movement of ions, this is only one of a wide variety of activities associated with membrane structures. Membranes are involved in functions such as cell–cell interactions, contractile processes, excitation in nerves, ATP generation, and energy conversion in photosynthesis. Many of these processes are in turn closely related to the movement of ions across membranes.

Composition of membranes

Our knowledge of membrane structure and composition has increased enormously in the last 20 years with the development of electron microscopy and modern biochemical techniques. However, it should be remembered that much was learnt of the nature of membranes long before they could readily be distinguished by electron microscopy. The presence of a cell membrane or plasma membrane was inferred by the early light microscopists from observations of the swelling and contraction of cells, the uptake of dyes, and the loss of contents when the cell surface is torn. Electron microscopy confirmed the universal existence of this membrane. However, even today, the selective role of the plasma membrane, particularly in relation to ions, is doubted by a minority of authors who claim that selectivity is a property of the cytoplasm. This view will be discussed more fully later (see p. 70).

One of the earliest predictions as to the nature of this membrane came from permeability studies of plant cells conducted by Overton towards the end of the nineteenth century. He demonstrated that the rate of penetration of non-electrolytes was correlated with their relative solubility in fats and so concluded that the plasma membrane was essentially lipoidal in nature. These observations were strengthened by the classic studies of Collander some 40 years later on the permeability of giant algal cells (Fig. 1.2). An approximately linear relationship was found between the oil solubility of various solutes and a parameter $PM^{1.5}$ which combines the permeability and molecular size of the solute. This relationship is discussed more fully in Chapter 3. Further evidence for the lipoidal nature of cell membranes came from studies of their electrical properties (e.g. capacitance, resistance), which were similar to these of lipid layers. However this was not the whole story. In 1935 Danielli and Davson reported that the surface tension of cells was much lower than the high values obtained for neutral lipid–water interfaces, suggesting that protein was present at the membrane

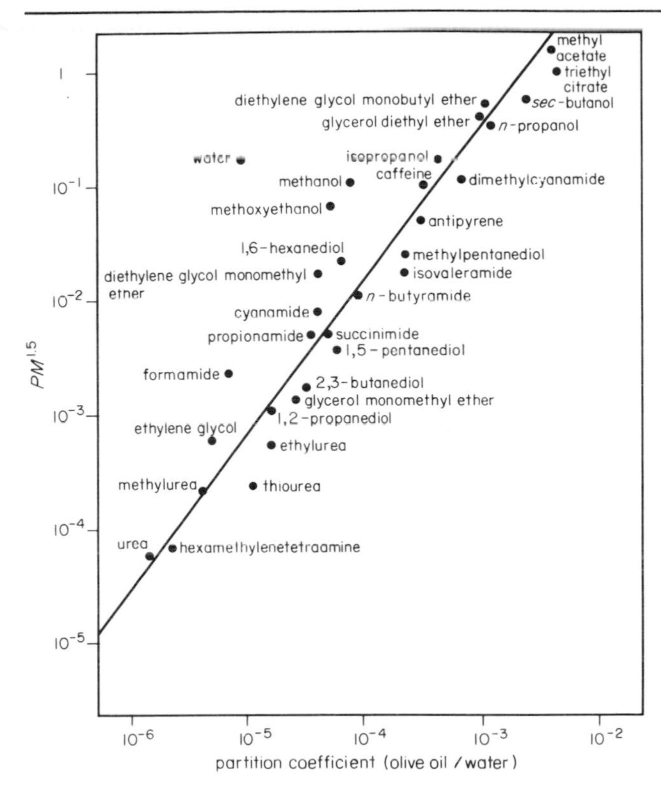

Fig. 1.2 Relationship between the permeability of several non-electrolytes towards cells of the giant alga *Nitella mucronata* and their relative oil solubility. P = permeability constant, M = molecular weight of solute. The value $PM^{1.5}$, which introduces a molecular size factor into the relationship, gave a better approximation to linearity than when permeability alone was considered. Note that there are a number of anomalous molecules (e.g. water, methanol, formamide) but these are of small size. This perhaps suggests the presence of very small pores in the membrane. (Redrawn from Collander, 1954.)

surface. The amphoteric nature of cell surfaces, which is also a characteristic property of protein molecules, and the fact that proteases could destroy cell membranes lent further support to the concept that protein was an essential component of cell membranes. Subsequent analysis of certain membranes isolated free from other cellular materials has shown that they consist largely of lipid and protein, together with some carbohydrate, thus confirming the predictions based on physiological and physical measurements. The carbohydrate is in the form of sugar residues attached to protein or lipid to form glycoproteins and glycolipids.

The chemical composition of membranes is not easy to determine since it is difficult to isolate them free from cytoplasmic contaminants, although with improving techniques of cell fractionation it has been possible to isolate a number of specific membrane fractions. However, it must be remembered that components may be lost or gained during the isolation procedure, and the danger of contamination by other cell membranes must never be overlooked. In addition, it must be realized that there is considerable heterogeneity among membranes. Thus a particular membrane fraction from a complex tissue may well vary with the different cell types within that tissue and even with the stage of growth and division of that cell type.

With these limitations in mind, the overall composition of a number of membrane types is shown in Table 1.2. It can be seen that there is considerable variation in the proportion of protein to lipid, ranging from 0.23

Table 1.2 Chemical composition of various cell membranes.

Membrane	Protein (%)	Lipid (%)	Carbohydrate (%)	Protein/lipid ratio
Myelin	18	79	3	0.23
Plasma membranes				
Blood platelets	33–42	58–51	7.5	0.7
Mouse liver cells	46	54	2–4	0.85
Human erythrocyte	49	43	8	1.1
Amoeba	54	42	4	1.3
HeLa cells	60	40	2.4	1.5
Nuclear membrane of rat liver cells	59	35	2.9	1.6
Mitochondrial outer membrane	52	48	(2–4)*	1.1
Sarcoplasmic reticulum	67	33		2.0
Chloroplast lamellae, spinach	70	30	(6)*	2.3
Mitochondrial inner membrane	76	24	(1–2)*	3.2
Gram-positive bacteria	75	25	(10)*	3.0
Halobacterium purple membrane	75	25		3.0

* Deduced from the analyses.
(Based on data collected by Guidotti, 1972.)

in nerve myelin to 3.0 in some bacterial and mitochondrial membranes. This variation must be taken into account when considering any general model of membrane structure. It is probable that the ratio of protein to lipid increases in relation to the functional and biochemical complexity of the membrane. Thus myelin, which acts as an inert insulator around nerve axons, has a low protein content and contrasts with the inner mitochondrial membrane, which is very active enzymatically and has a high protein content. Guidotti (1972) has suggested that membranes may be classified into three broad groups according to their gross chemical composition:

1. Simple membranes such as myelin which are composed largely of lipids and function mainly as insulators with little biochemical activity.
2. Membranes composed of about 50% protein, typified by animal plasma membranes. Compared with myelin, they have increased enzymatic and transport activities.
3. Membranes with protein as the major component, such as bacterial cell membranes and the inner mitochondrial membrane. They contain complex enzyme systems and are involved in processes such as transport, oxidative phosphorylation and the synthesis of nucleic acids.

Membrane lipids

As we have seen above lipids make up 25–80% by weight of cell membranes and the range of lipids varies widely among membranes from different sources. These lipids are usually *amphipathic*, i.e. they contain both hydrophobic (non-polar) and hydrophilic (polar) regions, and so, as we shall see later, readily form surface monolayers when spread on water.

The most common lipids found in eukaryotic membranes are phospholipids, glycolipids, sphingolipids, and sterols, the last two groups being largely absent from prokaryotic cells. The structures of some of these common membrane lipids are shown in Fig. 1.3. The long chain of methylene residues found in the fatty acids or in sphingosine form the hydrophobic regions of phospholipids, glycolipids or sphingolipids. The substitutions on the 3′ position of glycerol or on the head of the sphingosine (e.g. phosphoric acid derivatives, polar sugar groups) provide the hydrophilic regions. Sterols are based on the perhydrocyclopentanophenanthrene nucleus but show various degrees of unsaturation and substitution. One of these, cholesterol, is a common constituent of animal membranes; it is a neutral lipid with no polar region and is probably associated with the hydrophobic core of the membrane, inserted between phospholipid molecules. The lipids form a very heterogeneous group of compounds and are not easy to subdivide. For example, glycerolipids have glycerol as a structural backbone and include many phospholipids and glycolipids. However, certain phospholipids and glycolipids (which contain phosphoric acid and carbohydrate groups, respectively) are grouped in the sphingolipids since they contain a long-chain amino alcohol, sphingosine, as a structural backbone.

The proportions of these lipids in different membrane types varies widely, as shown in Fig. 1.4, although the significance of these differences in relation to membrane function is not clear. Some interesting observations have come from studies with red blood cells in which major changes

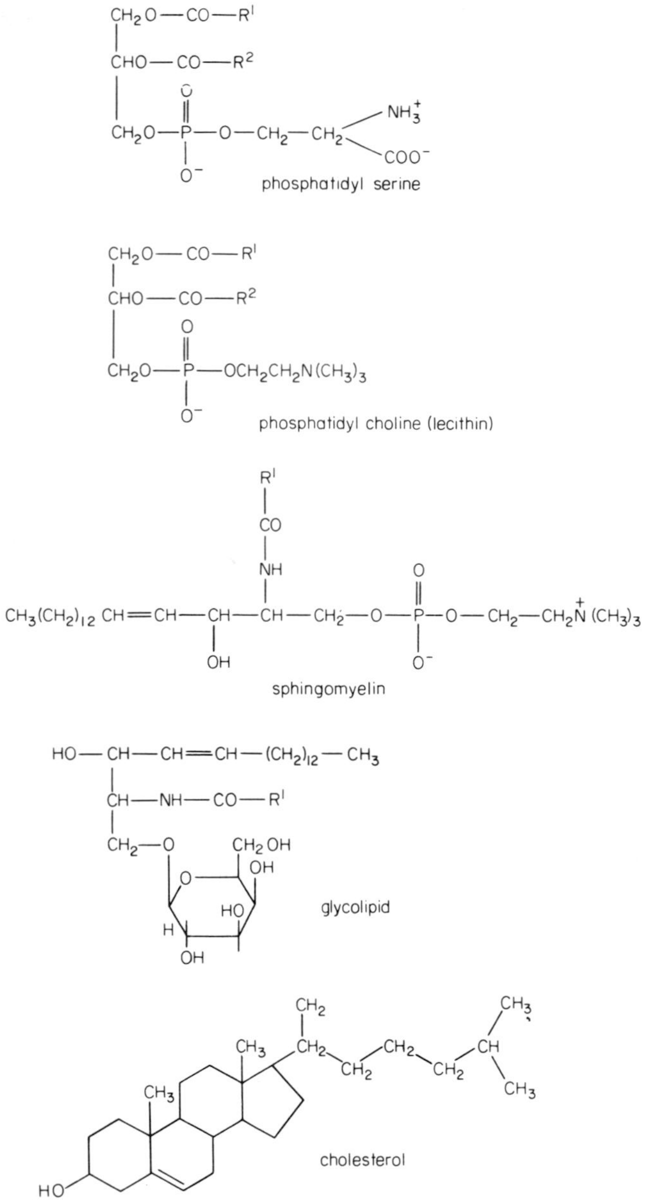

$CH_2O—CO—R^1$

$CHO—CO—R^2$

$CH_2O—P(=O)(O^-)—O—CH_2—CH_2(NH_3^+)(COO^-)$

phosphatidyl serine

$CH_2O—CO—R^1$

$CHO—CO—R^2$

$CH_2O—P(=O)(O^-)—OCH_2CH_2N(CH_3)_3$

phosphatidyl choline (lecithin)

R^1

CO

NH

$CH_3(CH_2)_{12}\ CH{=}CH—CH(OH)—CH—CH_2—O—P(=O)(O^-)—O—CH_2—CH_2\overset{+}{N}(CH_3)_3$

sphingomyelin

$HO—CH—CH{=}CH—(CH_2)_{12}—CH_3$

$CH—NH—CO—R^1$

$CH_2—O$

CH_2OH

OH

O

HO

H

OH

glycolipid

CH_2

CH_3

CH_3

CH_2

CH_2

CH_2

CH_2

CH

CH_3

CH_3

HO

cholesterol

Fig. 1.3 Structures of some common membrane lipids. R_1 and R_2 represent fatty acids. Phosphatidyl serine and phosphatidyl choline are phospholipids and contain serine and choline, respectively, at their polar heads. Sphingomyelin is a sphingolipid with a long chain amino alcohol as a structural backbone rather than glycerol; it contains choline esterified to the sphingosine phosphate. Glycolipids contain a sugar group and have a structured backbone of glycerol or sphingosine; the latter may be classed as sphingolipids.

---

in lipid composition are produced by dietary changes. A reversible depletion of cholesterol produces a decrease in cell surface area and an increase in osmotic fragility. It is known that cholesterol influences the packing of phospholipid in membranes, and so its depletion may result in changes to permeability or mechanical changes, or perhaps both. The phospholipids themselves vary widely from membrane to membrane (see Fig. 1.4) and may play a major role in membrane permeability and selectivity. For example, vesicles can be produced from phospholipids and, when prepared from phosphatidylserine and phosphatidylglycerol, it has been reported that they show a marked selectivity for potassium over sodium. Again, Ferguson and Simon (1973) have studied electrolyte loss and

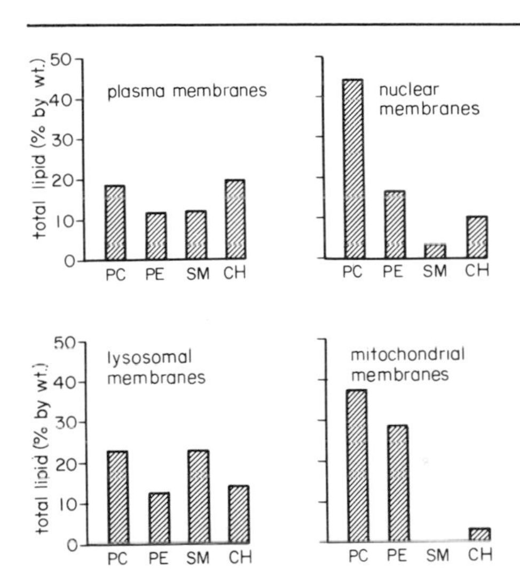

Fig. 1.4 Proportions of some major lipids in different membranes isolated from mammalian cells. PC, phosphatidyl choline; PE, phosphatidyl ethanolamine; SM, sphingomyelin; CH, cholesterol. (Based on data from van Hoeven and Emmelot, 1972.)

---

phospholipid content of senescing cucumber cotyledons and have attributed the increase in leakiness of the cells to a loss of phospholipid.

A knowledge of the behaviour of the amphipathic lipids in aqueous solutions is essential to an understanding of their arrangement in biological membranes. In aqueous systems polar lipids disperse to form globular

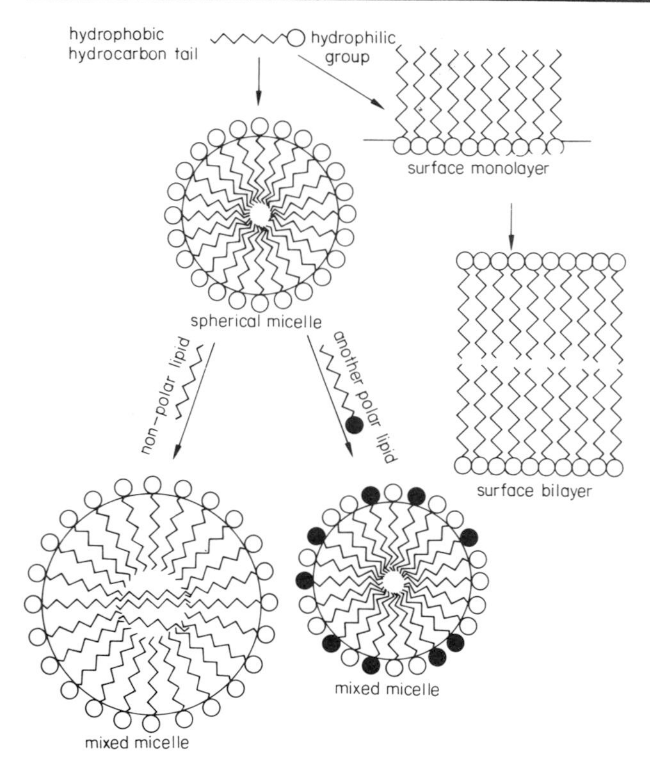

Fig. 1.5 The formation of micelles of polar, lipid molecules. The spherical micelles are shown in cross-section and for simplicity the lipids are drawn with only a single hydrophobic tail. (Reproduced from Hall, Flowers and Roberts, 1974.)

structures called *micelles*, in which the non-polar hydrocarbon tails are hidden from the aqueous environment and form a hydrophobic phase, while the hydrophilic heads are exposed at the surface and interact with water (Fig. 1.5). Thus phospholipids will disperse readily in aqueous solution though they do not form a truly molecular solution. Alternatively, polar lipids may interact with water to form a surface monolayer with the polar regions facing the aqueous phase (Fig. 1.5). If such a lipid monolayer is compressed, folding occurs and bilayers are produced (Fig. 1.5). These artificial bilayers are now used extensively in membrane research and, as we shall see later, many of their properties closely resemble those of natural membranes. Such bilayers may be prepared by sonicating phospholipid in water to produce lipid bilayer vesicles. An alternative method is to place a drop of lipid in solvent on a circular hole in a plastic film. As the solvent retracts to the edges of the hole, a sheet of bilayer lipid is left

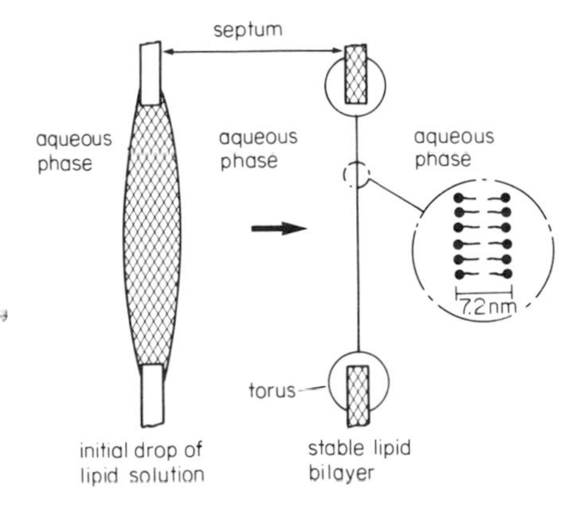

Fig. 1.6 Diagrammatic representation of the formation of an artificial lipid bilayer. The membrane-forming solution applied to the aperture thins spontaneously to a thickness of about 100 nm. At this point localized regions change to a much thinner structure of about 7.5 nm. These thinner regions increase rapidly until they fill the whole aperture. The excess lipid enters a torus at the edge of the aperture. (Redrawn from Thompson and Henn, 1970.)

spanning the orifice; this is called a *black lipid film* since it is black to reflected light (Fig. 1.6).

Membrane proteins

Whereas the membrane lipids can be extracted by organic solvents and so purified and analysed, membrane proteins have proved to be more difficult to study. They are insoluble under normal circumstances and are therefore difficult to dissociate from the membrane. However, great improvements have now been made in techniques for the extraction and solubilization of these proteins by organic solvents and detergents such as sodium dodecyl sulphate (SDS), and for their subsequent separation by centrifugation, chromatography and electrophoresis.

Such attempts to dissociate proteins from the membrane have led to the conclusion that there are probably two broad classes of membrane protein (Guidotti, 1972; Singer, 1974):

1. 'Peripheral' or weakly bound proteins which can be extracted by dilute EDTA solutions, strong salt solutions or sonication.
2. 'Integral' or tightly bound proteins which are dissociated from the membrane only by detergents or solvents.

Proteins of the first group are probably adsorbed onto the surface of the membrane lipid, whereas those of the second group are presumably firmly

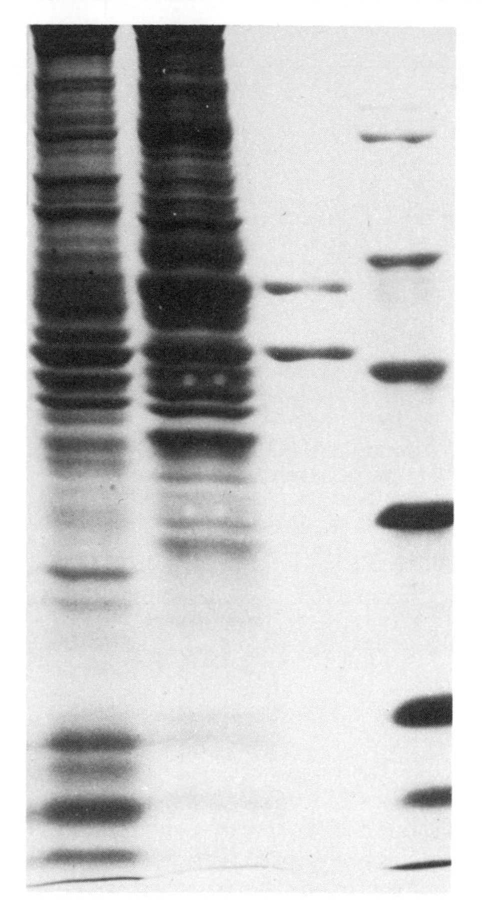

Fig. 1.7 Separation of plasma membrane proteins dissolved in sodium dodecyl sulphate by polyacrylamide electrophoresis. The markers consist of purified proteins of known molecular weight and include actin and tubulin which are recognized constituents of many plasma membranes and neuronal membranes, respectively. (Provided by Dr W. H. Evans, National Institute for Medical Research, Mill Hill, UK.)

embedded in the lipid. This conclusion is supported by analyses of the amino acid composition of these proteins. The second group often contains a higher proportion of hydrophobic residues, a finding which presumably is related to the penetration of these proteins in the lipid layer.

There appears to be a wide variation in the molecular weights of membrane proteins; this can be illustrated by the separation of membrane polypeptides by acrylamide gel electrophoresis using SDS to solubilize the proteins. The proteins are separated on the basis of their molecular weight and can be identified by staining with Coomassie brilliant blue or amido black (see Fig. 1.7). This method also allows an estimation to be made of the protein molecular weights, those from membranes usually ranging from about 20,000 to over 200,000. Some membranes, such as myelin, contain only a few proteins, while the mammalian red blood cell membrane contains at least 20, many of which will have enzymic activity.

In relation to ion transport, an interesting group of proteins have been isolated from bacteria. Bacteria possess a wide range of transport systems for amino acids, sugars and ions. Like enzymes, these are highly specific, suggesting an important role for proteins in these processes. If Gram-negative bacteria are subjected to osmotic shock, a set of proteins is released, known as *binding proteins*, each of which is able specifically to bind an ion, amino acid or sugar (see Oxender, 1972 and Singer, 1974, for a detailed discussion of these proteins). It is not clear how these proteins are bound to the membrane, but they are assumed to be peripheral proteins on the outer surface of the plasma membrane. In relation to ions, proteins which bind sulphate, phosphate and calcium have now been isolated and have molecular weights of approximately 32,000, 42,000 and 25,000–28,000, respectively. It seems most likely that these proteins are involved in solute recognition at the membrane surface, and this topic will be discussed in relation to the movement of ions across membranes in Chapter 3. Their relevance to ion transport studies in plants and animals is not known.

Membrane carbohydrates

Most cell membranes contain carbohydrates, either in the form of glycoproteins or glycolipids. Animal cell plasma membranes have been the most widely studied in this respect; the red cell membrane has received particular attention. The most common monosaccharides found in these plasma membranes are glucose, galactose, *N*-acetylglucosamine, *N*-acetylgalactosamine, mannose, fucose, neuraminic acid and sialic acid. These are thought to form branched oligosaccharides and are linked covalently to the protein or lipid.

The structure of glycolipids has been described on p. 8. Less is known of membrane glycoproteins, although the major glycoprotein of human erythrocyte membranes has been characterized. About 64% of the molecule is carbohydrate, mostly sialic acid together with galactose, mannose, fucose, acetylglucosamine and galactosamine. Some of these occur as tetrasaccharides with two sialic acids, an *N*-acetylgalactosamine and a galactose. The protein contains a single polypeptide chain with a total molecular weight of about 55,000.

The available evidence suggests that the carbohydrate of plasma membrane glycoproteins is located on the outer surface of the membrane. If whole cells are treated with the enzyme neuraminidase, sialic acid is liberated from the surface. Incubation of isolated membranes with colloidal iron hydroxide, which binds to acidic groups, stains only the exterior surface of the membrane. Both the glycoproteins and glycolipids are thought to have important roles in immunological responses, cell–cell adhesion and cell surface transformation in animal cells. Much less is known of the glycoprotein and glycolipid components of plant cell membranes.

Membrane structure

The observations described above show that cellular membranes are composed largely of proteins and lipids. What we have to consider now is how these molecules are arranged within the plane of the membrane. Many models have been proposed over the last 50 years aimed at explaining the structure of membranes on the basis of current observations, and it is interesting to follow the development of the study of membrane structure.

In 1925, Gorter and Grendel calculated that the area occupied by the lipid extracted from erythrocyte membranes was sufficient to form a layer two molecules deep – a bilayer – over the whole cell surface with the polar groups facing to the outside of the bilayer. It has subsequently been shown that these workers extracted only 70–80% of the total lipid and underestimated the surface area, so that the value of 2:1 obtained for the ratio of lipid surface area to cell surface area was obtained fortuitously due to self-cancelling errors. More recent and more accurate determinations have given a ratio of 1.3 to 2.2 depending on the state of compression of the lipids. However, the view that membranes consisted of lipid bilayers gained support until it was discovered that biological membranes have a much lower surface tension than that displayed by oil or neutral lipid–water interfaces. This suggested that protein was present at the cell surface, although it is now known that synthetic phospholipid bilayers have similar low surface tensions. However this led Danielli and Davson (1935) to propose that a stable membrane structure would be produced by a symmetrical arrangement of a lipid bilayer covered on either side by a layer of protein (Fig. 1.8*a*).

With the development of the electron microscope it became possible to examine membranes directly and a new era in the study of membranes began. It was found that, when fixed and stained in potassium permanganate or osmium tetroxide and examined in thin sections, a wide range of membranes from different cells and organelles showed a very similar image and dimension in the electron microscope. They are about 7.5 ± 2.00 nm thick and have a three-layered appearance consisting of two electron-dense lines separated by an unstained space (Fig. 1.9). This characteristic appearance of biological membranes led Robertson (1959) to propose a structural model, modified from that of Danielli and Davson, which he termed the

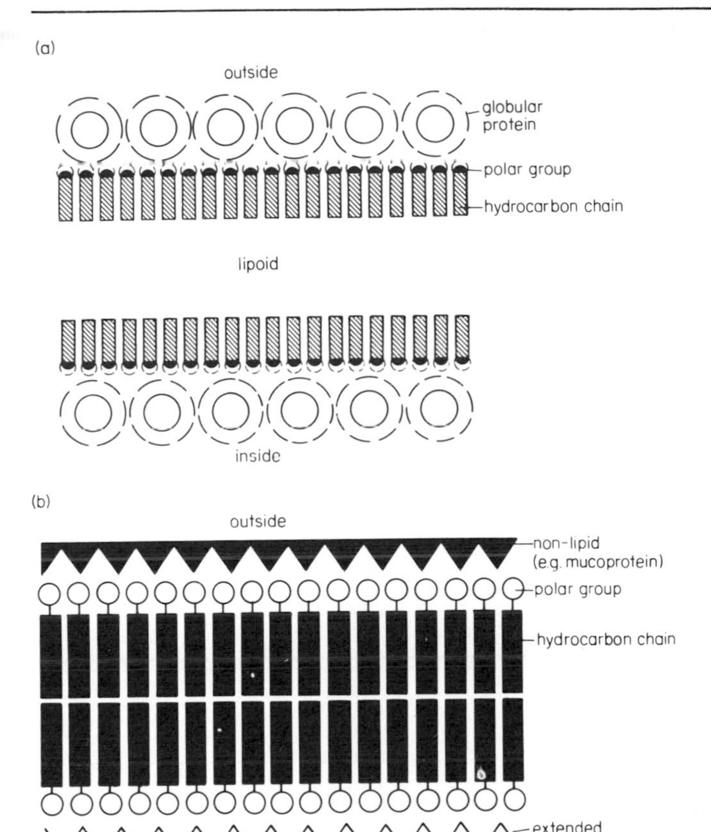

Fig. 1.8 Models of membrane structure. (*a*) According to Danielli and Davson (1935). (*b*) Unit membrane model of Robertson (1967).

unit membrane. This is shown in Fig. 1.8*b* and is considered to represent a universal structure for all biological membranes. It differs from the Danielli–Davson model in two respects. The proteins are considered to be in the extended β-form rather than globular, and allowance was made for possible differences between the inner and outer protein layers in, for example, glycoprotein content.

Since its proposal, evidence both for and against this generalized concept has been produced. This has been thoroughly reviewed by Hendler (1971) and will not be discussed in detail here. One serious objection to such a universal model comes from the wide variation found in the thickness, composition and activity of different membranes. Thus the thickness of

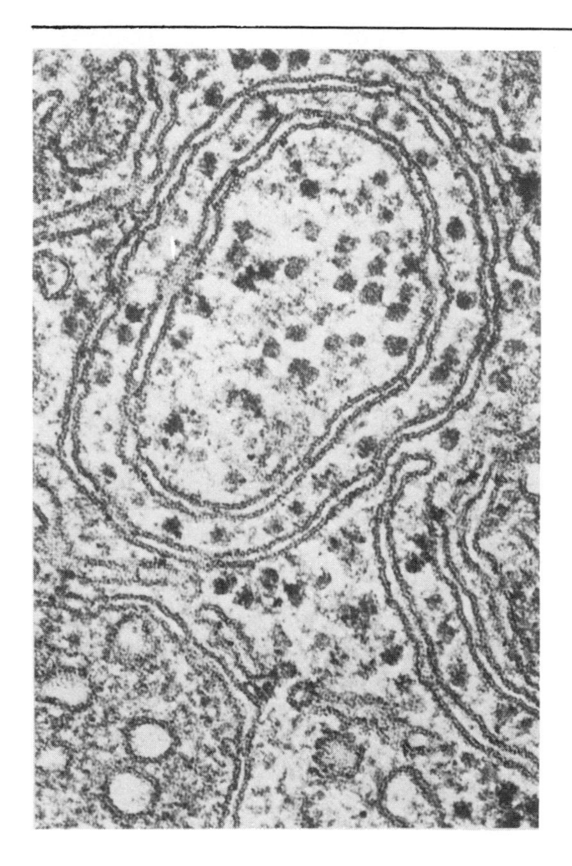

Fig. 1.9 Electron micrograph of part of a cell from the zone fasciculata of rat adrenal gland fixed with glutaraldehyde and osmium tetroxide. The characteristic unit membrane structure is seen in the membranes of the endoplasmic reticulum, × 200,000. (Provided by Dr Gillian R. Bullock, CIBA Laboratories, Horsham, UK.)

plasma membranes is normally about 9.0 nm, while that of mitochondrial membranes is between 5.0–6.0 nm. In addition, as we have seen earlier, there is great variation in the activity and composition (e.g. protein/lipid ratio, phospholipid content) between membranes. It was pointed out that little is known of the reactions involved in fixing or staining membranes or what structural alterations these procedures might produce in the membrane. It is interesting that the removal of 95% of the lipid from the inner mitochondrial membrane does not alter its apprearance when fixed in osmium and viewed by electron microscopy. Finally the technique of freeze-etching, which will be discussed in detail later, has revealed the presence of globules or particles within the membrane matrix, an observation which is difficult to reconcile with the unit membrane model.

These objections to the model proposed by Robertson have given rise to a series of alternative proposals to explain the arrangement of protein and lipid. The lipid bilayer theory has not been accepted by everyone, and it has been suggested that the lipids might be arranged as globular micelles embedded in a protein matrix. However, most authorities have concluded that a lipid bilayer forms a substantial part of biological membranes and is the most satisfactory explanation of its unspecialized properties. Perhaps the best evidence for this is provided by studies of the properties of artificial bilayers and by X-ray diffraction analysis of packed membrane preparations.

The preparation of artificial lipid bilayers has been discussed before (p. 10) and, as with natural membranes, they contain a thin hydrophobic interior. Thus it is of interest to compare the electrical and permeability properties of these two systems, as shown in Table 1.3, together with other features. It can be seen that the thickness, permeability to water and small molecules, surface tension and capacitance are very similar. However, other properties such as electrical resistance and ionic permeability are quite different, showing that biological membranes are not simply continuous lipid bilayers; presumably, in natural membranes, the bilayer properties are modified by the presence of protein. Artificial membranes may be made to resemble natural membranes in terms of ionic permeability by the addition of ionophores (see p. 65) and the significance of this to ion transport studies is discussed in Chapter 3.

Table 1.3 Comparison of the physical properties of natural membranes and artificial lipid bilayers.

Property	Natural membrane	Artificial membrane
Thickness (nm)	5–10	6.5–7.3
Water permeability (10^{-6} m s^{-1})	0.4–400	5–10
Urea permeability (10^{-8} m s^{-1})	0.02–280	4.2
Surface tension (μN m^{-1})	0.003–0.3	0.05–0.2
Capacitance (10^4 μF m^{-2})	0.5–1.3	0.4–1.3
Resistance (Ω m^{-2})	10^7–10^{10}	10^{10}–10^{13}

X-ray diffraction is an important method in the study of the structure of molecules and has produced useful information in relation to membrane structure. X-rays are reflected by atoms; passing a beam of X-rays through a crystal produces a characteristic reflection pattern on a photographic plate which enables the arrangement of atoms in the crystal to be deduced.

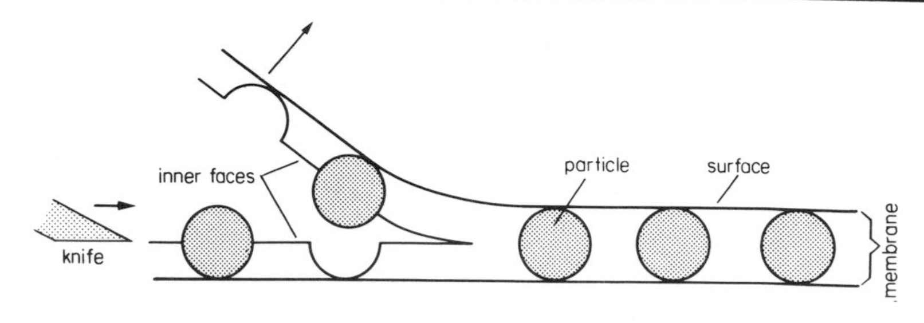

Fig. 1.12 Diagrammatic representation of the plane of fracture produced during freeze-etching of a membrane.

tant constituent; it was pointed out earlier that there is also a correlation between metabolic activity and the protein content of membranes.

Such evidence suggests that there may be considerable penetration of the lipid bilayer by proteins, and this is supported by some elegant labelling experiments carried out with erythrocytes. Bretscher (1971) showed that when intact cells were treated with formylmethionyl sulphone methyl phosphate, a reagent which binds with amino groups but which does not penetrate the membrane, parts of two membrane proteins were labelled with this reagent. Presumably these were the parts exposed at the outer sur-

Table 1.4 Particle densities on freeze-fractured membrane faces.

	Number of particles per μm^2 of membrane face		
Membrane	**Densely populated face**	**Thinly populated face**	**Membrane area covered by particles (%)**
Nerve myelin	0	0	0
Nuclear membrane (root tip)	1,790	420	12
Plasma membrane (root tip)	2,030	550	15
Plasma membrane (red blood cell)	2,800	1,400	23
Mitochondrial inner membrane	2,700	—	—
Vacuole (root tip)	3,300	2,480	32
Plasma membrane (yeast)	2,600	—	63
Chloroplast lamellae	3,860	1,800	80

(Based on data collected by Branton, 1969.)

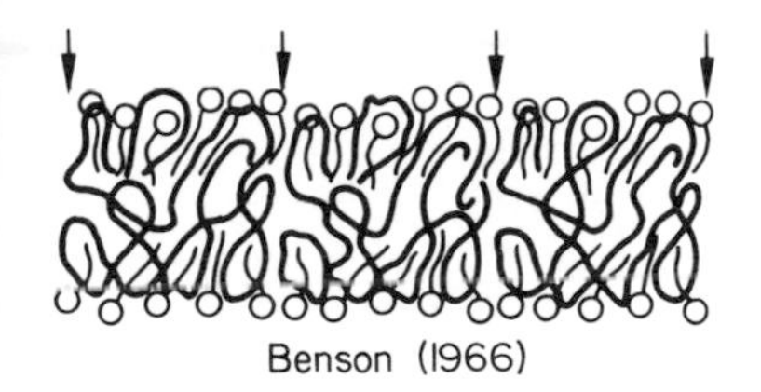

Benson (1966)

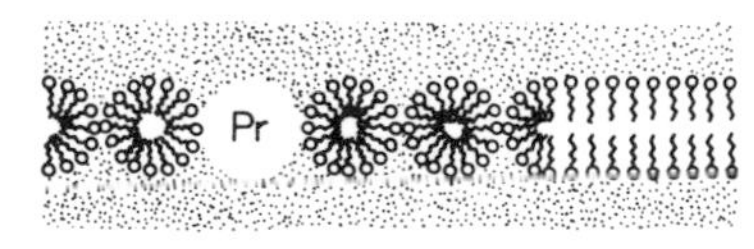

Lucy (1968)

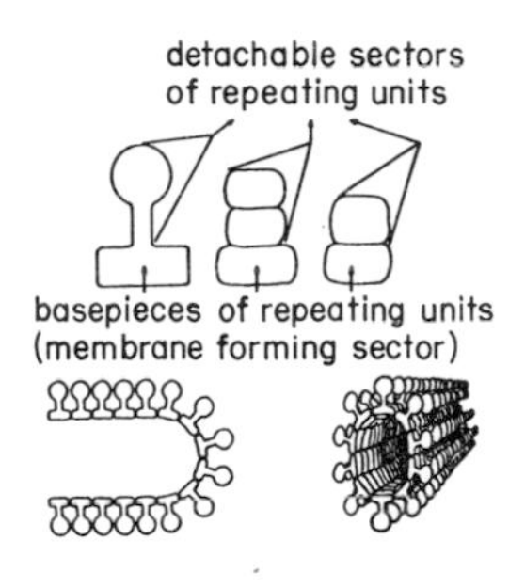

Green and Perdue (1966)

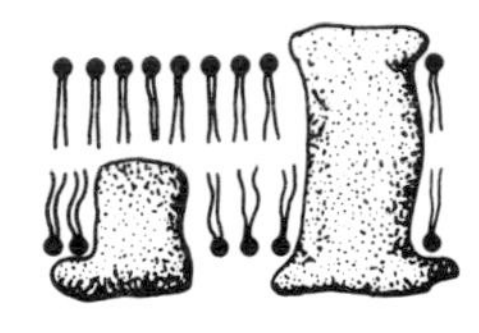

Wallach and Zahler (1966)

Fig. 1.13 Models of membrane structure. Benson (1966), a series of lipoprotein subunits; Lucy (1968), lipid globules and bilayers embedded in a protein matrix; Green and Perdue (1966), different forms of repeating lipoprotein subunits and their arrangement in the membrane; Wallach and Zahler (1966), hydrophilic regions of membrane polypeptides located at both surfaces and connected by hydrophobic zones.

face of the cell. However, when 'ghosts' (prepared by osmotic lysis of erythrocytes making both sides of the membrane accessible to the reagent) are treated, other parts of these proteins become labelled. It was concluded that these particular proteins traverse the lipid bilayer with parts exposed at both the inner and outer surfaces of the membrane. Further labelling experiments, which cannot be described in detail here, have demonstrated that lateral mobility of proteins in the plane of the membrane may also occur. The proteins have been considered as 'floating' in a fluid lipid matrix.

Such observations clearly raised considerable doubts as to the validity of the unit membrane concept, which makes no allowance for such penetration and mobility of the membrane proteins. The result has been the proposal of a wide range of models of membrane structure some of which are shown in Fig. 1.13. The two extremes are represented by the unit membrane model and that proposed by Benson (1966), which consists of a series of lipoprotein subunits, each with the hydrocarbon tails of the lipid

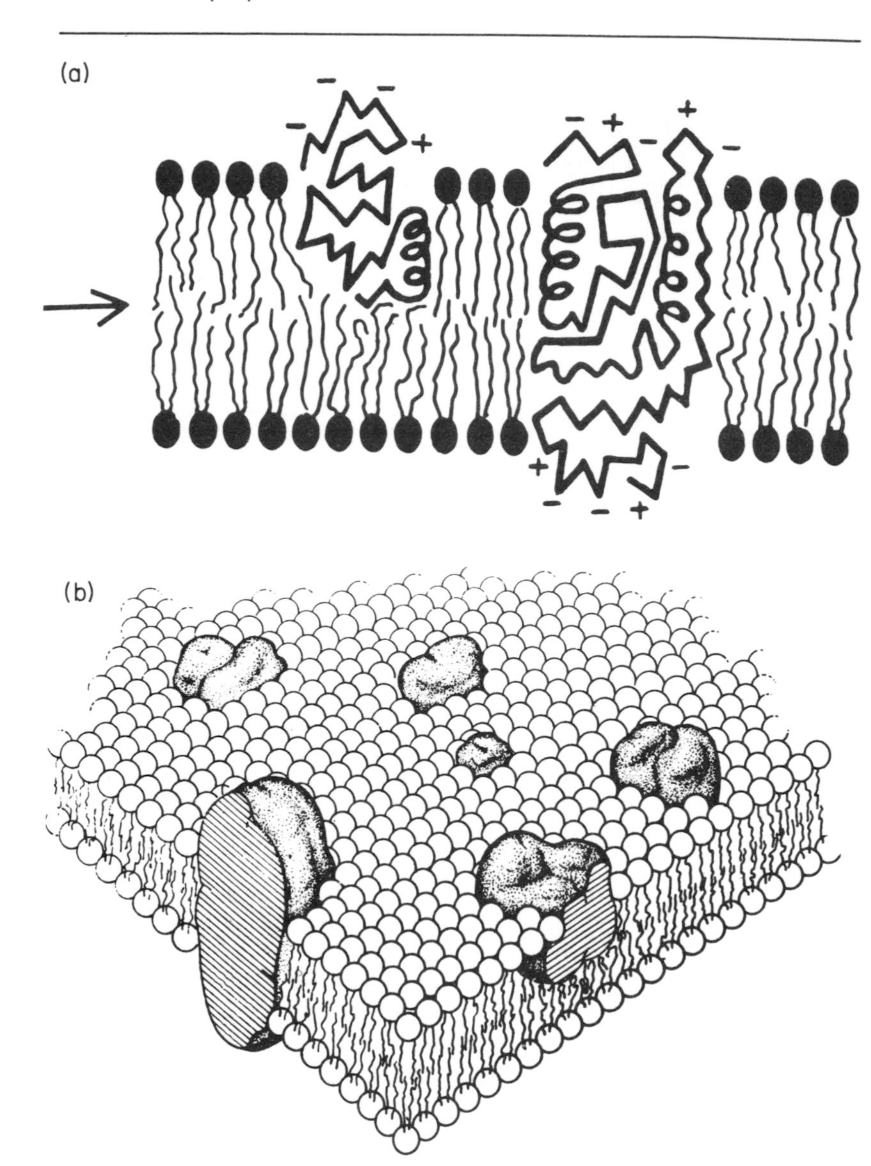

Fig. 1.14 The lipid–globular protein mosaic model of membrane structure according to Singer and Nicolson (1972). (*a*) Cross-sectional view. The phospholipids are arranged in a discontinuous bilayer. The proteins are shown as globular molecules, partially embedded in the membrane. The arrow marks the plane of cleavage to be expected in freeze-etching. (*b*) Three-dimensional view. The solid bodies with stippled surfaces represent the globular integral proteins.

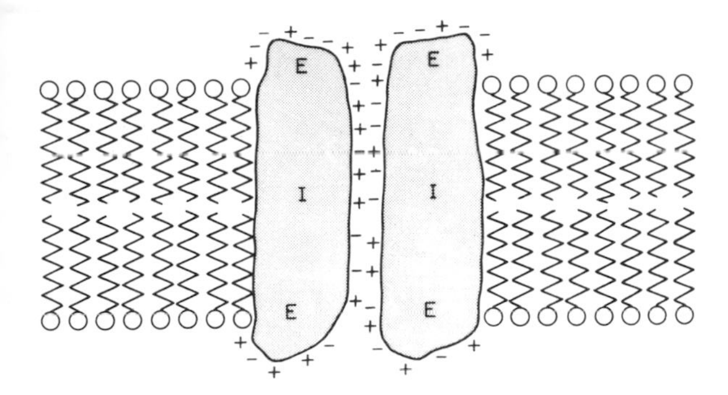

Fig 1.15 Diagrammatic representation of possible arrangement of integral proteins as subunit aggregates which span the membrane and generate water-filled pores through the membrane. E and I refer to exterior and interior regions of the protein, respectively. (Redrawn after Singer, 1974.)

inserted in a folded protein chain. However, most models include a lipid bilayer which is thought to form a structural framework to which proteins are attached and embedded to provide the functional centres within the membrane. Many of these ideas have been brought together by Singer and Nicolson (1972) in the *fluid mosaic model.* They consider that the basic structure of biological membranes is a two-dimensional arrangement of globular integral proteins dispersed in a matrix of fluid lipid bilayer (Fig. 1.14). The proteins are amphipathic molecules with their ionic and polar groups embedded in the hydrophobic interior of the membrane. This model does not include the peripheral proteins, which are presumably associated with the polar surface of the bilayer occupying the space between the integral proteins. It must be added that much of the evidence relating to membrane structure has come from the study of plasma membranes, and detailed studies of other membranes, although more difficult to undertake, are needed before the universality of such a model can be accepted.

Finally, while dealing with membrane structure, mention must be made of the occurrence of *membrane pores.* Evidence for the presence of pores has come from studies of membrane permeability. They have been proposed to account for the high permeability of the plasma membrane to water and some other small molecules; this evidence is discussed in detail in Chapter 3. Organized pores, however, have not been observed in the plasma membrane, although this is not surprising since their size (calculated minimum radius 0.4–0.5 nm) would not allow them to be visualized by standard techniques of electron microscopy. Furthermore, the pores might not exist as permanent channels but as transient, aqueous gaps formed by various molecular movements. For example, movement of the lipid components would mean continual changes in the distances between hydrocarbon chains, which could result in pore formation. Alternatively, if the integral proteins which span the membrane exist as subunit aggregates, water-filled pores may be formed (Fig. 1.15). However, good experimental support is still required for these proposals.

Membrane dynamics

The picture of membranes presented so far is perhaps one of static diffusion barriers separating the cell from its environment and dividing the cell internally. This impression of static structures is given by many electron micrographs. However, it is clear from other observations that membranes are dynamic, mobile structures. Numerous examples of this may be seen even without the aid of the electron microscope. Cells such as *Amoeba*, slime moulds and phagocytes show continual changes in shape and surface area, which implies considerable plasticity of the surface membrane. Membranes show a capacity for self-repair and so, if the cell surface is punctured, it can often be sealed quickly. The fusion of two cells may require considerable reorganization of the surface membrane. Also, plant cell protoplasts can be made to shrink and then expand by placing them in a solution of high concentration followed by one of low osmotic strength. These extensive changes may require disassembly or reassembly of the membrane, since there is little scope for shrinkage or expansion of the phospholipid bilayer structure without removing or adding membrane components.

This flexibility is an important property of the membrane in relation to certain transport processes known as *phagocytosis* and *pinocytosis*. These processes are collectively referred to as *endocytosis*, describing phenomena in which various materials are drawn into cells by means of invagination of the plasmalemma (Fig. 1.16). Such invaginations can readily be demonstrated by the use of an electron-dense, tracer molecule such as ferritin, the uptake of which may be followed by electron microscopy (Fig. 1.17). Phagocytosis generally refers to the engulfing of particulate matter (such as bacteria by larger cells), while pinocytosis involves the uptake of particulate-free extracellular fluid. The contribution of these phenomena to the transport of ions is not clear. They are generally considered as methods of bulk transport and are therefore unselective. However, it is possible to envisage a selective transport process involving pinocytosis in which invagination of the plasma membrane acts as a carrier mechanism for the movement of ions across this membrane. This is discussed further in Chapter 3. It is interesting to note that, in a study of pinocytosis in smooth muscle cells, Orci and Perrelet (1973) observed an increase in the density of membrane particles seen by freeze-etching in areas of the membrane involved in pinocytosis. The role of these particles, however, is not clear.

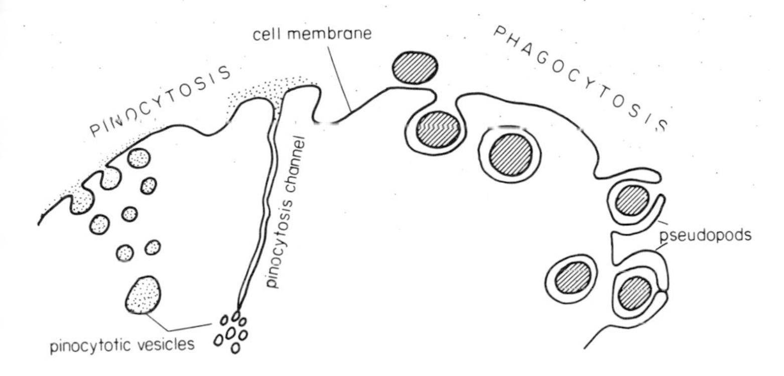

Fig. 1.16 Diagrammatic representation of forms of pinocytosis and phagocytosis. The formation of pinocytosis channels and use of pseudopods is characteristic of *Amoeba.*

Finally, in relation to the dynamic nature of membranes, some mention should be made of membrane turnover and membrane flow. *Membrane turnover* refers to the flow of constituents through or within a particular membrane and is a measure of the dynamic state of membranes. Even in tissues with very little growth or cell division there is still a steady turnover of the membrane components, both proteins and lipids. Not only is there turnover, but this frequently occurs at different rates relative to each other (Table 1.5). Siekevitz (1972) has likened membranes to an elastic, deformable brick wall, composed of bricks of many colours; each brick is being constantly replaced by a brick of similar colour, but these replacements are not in harmony with each other. Thus the wall, or membrane, persists, but its constituent molecules are continuously being replaced.

Another interesting concept is that of *membrane flow*, which involves the physical transfer of material from one cellular membrane to another. This concept is essential to the functioning of the *endomembrane system* as envisaged by Morré and Mollenhauer (1974); this is considered to be a continuum of internal cytoplasmic membranes but perhaps excluding the semi-autonomous organelles (mitochondria and chloroplasts). The endoplasmic reticulum (ER) is seen as a probable source of membrane building blocks. Membranes are transferred and transformed in a subcellular developmental pathway which involves the Golgi system and eventual incorporation into the plasma membrane (Fig. 1.18). An alternative route is to the interior of the cell where ER and Golgi system membranes may give rise to lysosomal or vacuolar membranes. Evidence for the operation of this system comes from a variety of fine-structural and biochemical observations. Examination of the ER, Golgi body membranes and plasma membrane shows a progressive increase in the dimensions of these membranes

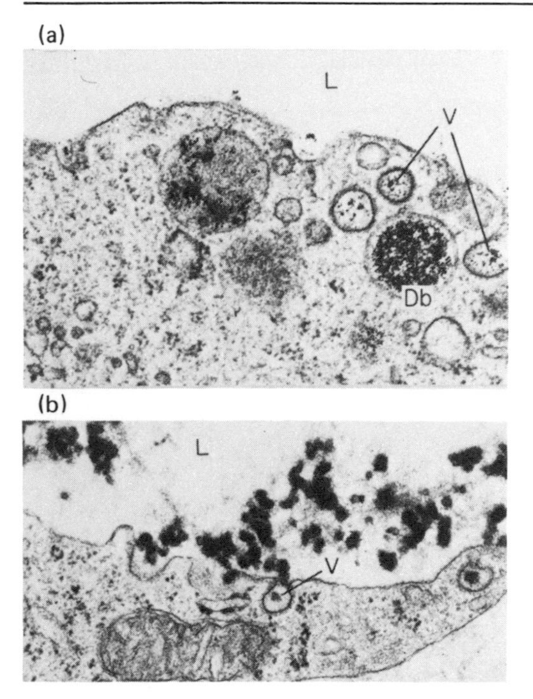

Fig. 1.17 Endocytosis in endothelial cells of rat bone marrow. Electron micrographs show the uptake of (*a*) ferritin (× 43,000) and (*b*) colloidal carbon (× 40,000) by plasma membrane vesicles. The particles may later accumulate in dense bodies. L, lumen; V, vesicles; Db, dense body. (From de Bruyn, Michelson and Becker, 1975.)

Table 1.5 Half lives of some membrane-associated components.

(*a*)	Enzymes of endoplasmic reticulum (rat liver)	$t_{1/2}$
	Hydroxymethylglutaryl CoA reductase	2–3 h
	Cytochrome *c* reductase	60–80 h
	Cytochrome b_5	100–120 h
	NAD glucohydrolase	approx. 380 h
(*b*)	Lipids of mitochondrial membranes	
	Phosphatidyl choline	2 weeks
	Phosphatidyl serine	3 weeks
	Phosphatidyl ethanolamine	4 weeks
	Sphingomyelin	1 month

((*a*) Taken from Schimke, 1975; (*b*) taken from O'Brien, 1967.)

and progressive changes in composition and enzyme activity (see Table 1.6), presumably associated with a steady gain or loss of components.

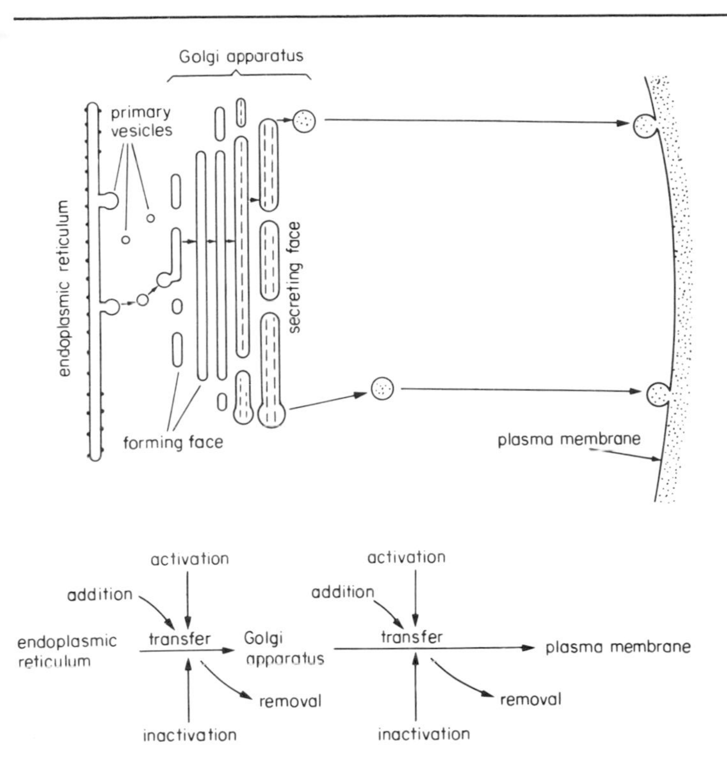

Fig. 1.18 Diagrammatic representation of the endomembrane concept involving membrane flow and differentiation. New membranes are synthesized at the rough endoplasmic reticulum (ER). Small primary vesicles arise from the ER and membranes and other materials are transferred to the forming face of the Golgi apparatus (GA). Secretory vesicles from the secreting face of the GA migrate to and fuse with the plasma membrane. Various types of transformation might account for the changes in membrane composition occurring during this flow. (Redrawn from Morré and Mollenhauer, 1974.)

Furthermore, if the rate of incorporation of amino acids is measured by injection of a radioactive tracer, a peak of incorporation is seen first in the ER and nuclear envelope membranes, then in the Golgi body and finally in the plasma membrane (Table 1.6). This finding is consistent with the labelling of the plasma membrane by a flow mechanism.

Another interesting example of the close interrelationships among membranes has been studied by Beevers and co-workers using the endosperm tissue of castor bean seeds (Kagawa, Lord and Beevers, 1973). With germination, fat breakdown is initiated and the numbers of glyoxysomes and mitochondria increase rapidly during the first 5 days. ^{14}C-choline was

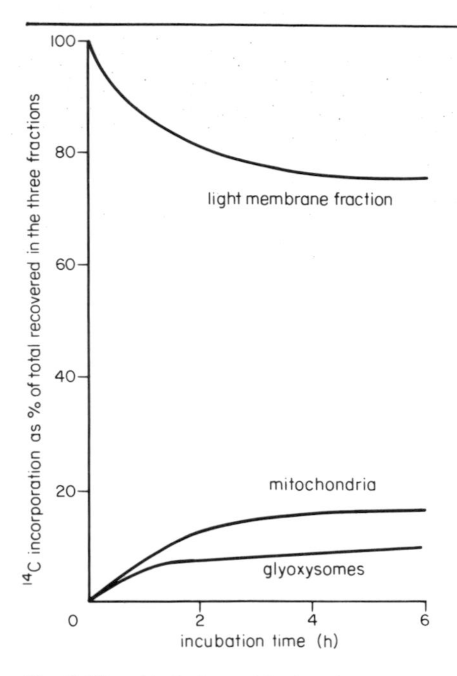

Fig. 1.19 Variation with time in the per cent distribution of total particulate ^{14}C in three membrane fractions isolated from 4-day-old castor bean endosperm after labelling with ^{14}C-choline. (Redrawn from Kagawa, Lord and Beevers, 1973.)

Table 1.6 Characteristics of membrane fractions from onion stem (*a*) and rat liver (*b*)–(*f*).

		Endoplasmic reticulum	**Golgi apparatus**	**Plasma membrane**
(*a*)	Average membrane thickness (nm)	4–5	5–7	9.5
(*b*)	Sialic acid (n mol mg^{-1} protein)	2.5	18	32
(*c*)	5′-nucleotidase activity (μ mol h^{-1} mg^{-1} protein)	1.9	3.2	54
(*d*)	Mg^{2+}-ATP-ase activity (μ mol h^{-1} mg^{-1} protein)	2	2.5	45
(*e*)	GDP-ase (μ mol h^{-1} mg^{-1} protein)	4.0	2.2	1.5
(*f*)	Time of half maximal labelling with ^{14}C-guanido-L-arginine (min)	5–7	10	15

(From data compiled by Morré and Mollenhauer, 1974.)

used as a precursor of membrane phospholipids, and its incorporation into mitochondria, glyoxysomes and a light membrane fraction (largely endoplasmic reticulum) was followed. All three fractions became labelled when ^{14}C-choline was applied to endosperm tissue but the light membrane fraction became labelled first. When incorporation is expressed as a percentage of the total recovered in the three fractions, it can be seen that ^{14}C is initially present almost entirely in the light membrane fraction but later appears in the mitochondria and glyoxysomes (Fig. 1.19). This would be expected if the light membrane fraction gives rise to the membranes of the two cell organelles.

In this chapter we have considered the composition, molecular arrangement and dynamic nature of biological membranes, which form the major barriers to the random diffusion of ions into and within plant and animal cells. The movement of ions across membranes is a controlled process involving specific transport mechanisms. Before we consider how this is achieved we must first discuss the nature of the forces which drive molecules across membranes.

Further reading and references

Further reading

BRETSCHER, M. S. and RAFF, M. C. (1975) Mammalian plasma membranes, *Nature*, **258**, 43.

GUIDOTTI, G. (1972) Membrane proteins, *Ann. Rev. Biochem.*, **41**, 731.

HENDLER, R. W. (1971) Biological membrane ultrastructure, *Physiol. Rev.*, **51**, 66.

MORRÉ, D. J. (1975) Membrane biogenesis, *Ann. Rev. Plant Physiol.*, **26**, 441.

MORRÉ, D. J. and MOLLENHAUER, H. H. (1974) The endomembrane concept: a functional integration of endoplasmic reticulum and Golgi apparatus, in *Dynamic Aspects of Plant Ultrastructure*, p. 84, ed. A. W. Robards. McGraw-Hill, London.

PARSONS, D. S. (ed.) (1975) *Biological Membranes: Twelve Essays on their Organization, Properties, and Functions*. Clarendon Press, Oxford.

SINGER, S. J. (1974) The molecular organization of membranes, *Ann. Rev. Biochem.*, **43**, 805.

SINGER, S. J. and NICOLSON, G. L. (1972) The fluid mosaic model of the structure of cell membranes, *Science*, **175**, 720.

WALLACH, D. F. H. (1972) *The Plasma Membrane: Dynamic Perspectives, Genetics and Pathology*. English Universities Press, London.

Other references

BENSON, A. A. (1966) On the orientation of lipids in chloroplast and cell membranes, *J. Am. Oil Chem. Soc.*, **48**, 265.

BRANTON, D. (1969) Membrane Structure, *Ann. Rev. Plant Physiol.*, **20**, 209.

BRETSCHER, M. S. (1971) Major human erthyrocyte glycoprotein spans the cell membrane, *Nature New Biol.*, **231**, 229.

COLLANDER, R. (1954) The permeability of *Nitella* cells to non-electrolytes, *Physiol. Plant,* **7**, 420.
CRAM, W. J. (1975) Storage tissues, in *Ion Transport in Plant Cells and Tissues,* p. 161, eds D. A. Baker and J. L. Hall. North Holland, Amsterdam and London.
DANIELLI, J. F. and DAVSON, H. (1935) A contribution to the theory of permeability of thin membranes, *J. cell. comp. Physiol.,* **5**, 495.
De BRUYN, P. P. H., MICHELSON, S. and BECKER, R. P. (1975) Endocytosis, transfer tubules, and lysosomal activity in myeloid sinusoidal endothelium, *J. Ultrastr. Res.,* **53**, 133.
FERGUSON, C. H. R. and SIMON, E. W. (1973) Membrane lipids in senescing green tissues, *J. exp. Bot.,* **79**, 307.
GREEN, D. E. and PERDUE, J. F. (1966) Membranes as expressions of repeating units, *Proc. natl. Acad. Sci. USA,* **55**, 1295.
HALL, J. L., FLOWERS, T. J. and ROBERTS, R. M. (1974) *Plant Cell Structure and Metabolism.* Longman, London.
HOLDEN, J. T. (1968) Evolution of transport systems, *J. theor. Biol.,* **21**, 97.
KAGAWA, T., LORD, J. M. and BEEVERS, H. (1973) The origin and turnover of organelle membranes in castor bean endosperm, *Plant Physiol.,* **51**, 61.
KOTYK, A and JANÁČEK, K. (1970) *Cell Membrane Transport: Principles and Perspectives.* Plenum Press, New York.
LUCY, J. A. (1968) Ultrastructure of membranes: micellar organization, *Brit. Med. Bull.,* **24**, 127.
NYSTROM, R. A. (1973) *Membrane Physiology.* Prentice-Hall, Englewood Cliffs, NJ.
O'BRIEN, J. S. (1967) Cell membranes – composition: ultrastructure: function, *J. theor. Biol.,* **15**, 307.
ORCI, L. and PERRELET, A. (1973) Membrane-associated particles: increase at sites of pinocytosis demonstrated by freeze-etching, *Science,* **181**, 868.
OXENDER, D. L. (1972) Membrane transport, *Ann. Rev. Biochem.,* **41**, 777.
PARDEE, A. R. and PALMER, L. M. (1973) Regulation of transport systems: a means of controlling metabolic rates, *Symp. Soc. Exp. Biol.,* **27**, 133.
ROBERTSON, J. D. (1959) The ultrastructure of cell membranes and their derivatives, *Biochem. Soc. Symp.,* **16**, 3.
ROBERTSON, J. D. (1967) *Molecular Organisation and Biological Function,* ed. J. M. Allen. Harper and Row, New York.
SADDLER, H. D. W. (1970) The ionic relations of *Acetabularia mediterranea, J. exp. Bot.,* **21**, 345.
SCHIMKE, R. T. (1975) Turnover of membrane proteins in animal tissues, in *Biochemistry of Cell Walls and Membranes,* Vol. 2, p. 229, ed. C. F. Fox, MTP International Review of Science, Butterworths, London.
SIEKEVITZ, P. (1972) Biological membranes: the dynamics of their organization, *Ann. Rev. Physiol.,* **34**, 117.
THOMPSON, T. E. and HENN, F. A. (1970) Experimental Phospholipid model membranes, in *Membranes of Mitochondria and Chloroplasts,* p. 1, ed. E. Racker. Van Nostrand Reinhold, New York.
van HOEVEN, R. P. and EMMELOT, P. (1972) Studies on plasma membranes XVIII. Lipid class composition of plasma membranes isolated from rat and mouse liver and hepatomas, *J. Membrane Biol.,* **9**, 105.
WALLACH, D. F. H. and ZAHLER, P. H. (1966) Protein conformations in cellular membranes, *Proc. natl. Acad. Sci. USA,* **56**, 1552.

Chapter 2

Ion transport: an analysis of the driving forces

Having considered the composition and structure of cell membranes we can now go on to consider the forces responsible for the movement of ions across them. Before we can interpret these driving forces in relation to living membranes it is necessary to consider the behaviour of ions in solution and the driving forces for the passage of ions across inert membrane barriers.

Diffusion

If a difference of concentration exists within a solution a randomization of solute molecules will take place, eventually resulting in an equal distribution throughout the system. This redistribution of particles occurs as a result of the kinetic energy of translation and rotation of the solute molecules and subsequent collisions between them, a greater number of molecules therefore moving from more to less concentrated regions than in the reverse direction. This process, *diffusion*, thereby brings about a random distribution of solute molecules within a solution. Although movement of solute molecules will continue after randomization, no net flow results and the system cannot return to its original condition unless work is done on it. Originally the solution had potential energy as a result of the concentration difference, but during the process of diffusion this free energy is dissipated as heat and increased entropy. Thus in thermodynamic terms the system has moved towards maximum entropy.

The rate of diffusion, $\mathrm{d}\nu/\mathrm{d}t$, is related to the concentration gradient, $\mathrm{d}c/\mathrm{d}x$, by *Fick's first law*:

$$\frac{\mathrm{d}\nu}{\mathrm{d}t} = -DA\,\frac{\mathrm{d}c}{\mathrm{d}x}, \qquad [2.1]$$

where D is the diffusion coefficient and A the area across which diffusion

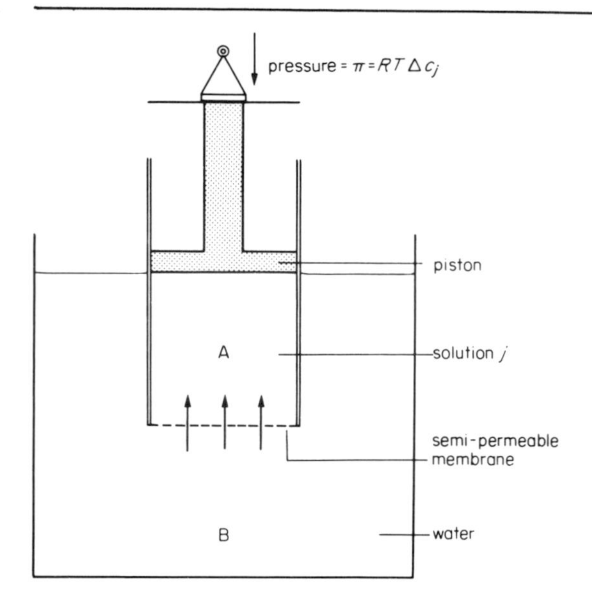

Fig. 2.1 A simple system for measuring π, the osmotic pressure. A cylinder A with a watertight sliding piston and a semi-permeable membrane containing a solution j is placed in pure water B. The pressure necessary to prevent movement of water from B to A is the osmotic pressure of the solution in A.

occurs. The negative sign is a convention to indicate that diffusion occurs from a higher to a lower concentration.

The force which causes the diffusion is related to the gradient of concentration and its resultant chemical activity by a function termed the chemical potential μ_j if the particle is uncharged and to the gradient of electrochemical potential $\bar{\mu}_j$ if the particle is electrically charged (see p. 34).

Osmosis

If a semi-permeable membrane, which permits solvent but not solute to move across it, is introduced into the above system, the potential energy of the concentration difference can be used to perform work. This situation is depicted in Fig. 2.1.

Taking water as the solvent, work will be performed by moving water by *osmosis*, a process in which a head of water is built up, the magnitude of which is determined by the original gradient of chemical potential. This movement of water could be prevented by applying a pressure to *A* (Fig. 2.1); the magnitude of the force needed to prevent movement of water into A is termed the *osmotic pressure* of solution A. The pressure required π is equal to $RT\Delta c_j$, where R is the gas constant, T temperature absolute and Δc_j the concentration gradient of solute j (or strictly its chemical potential gradient $\Delta\mu_j$).

Reflection coefficient

If, as is frequently the case, the membrane is not completely semi-permeable but permits the passage of some solute as well as solvent molecules, then the osmotic pressure required to prevent movement of water into A would be less and would not be directly equal to $RT\Delta c_j$. In this situation the *apparent osmotic pressure* π' is required, which may be defined as that pressure difference exerted across the membrane in order that there shall be no volume flow through the membrane. The ratio π'/π is termed the *reflection coefficient* σ, which gives an indication of the leakiness of the membrane to solute particles, i.e. the proportion of solute particles which are reflected (see p. 55). If the membrane is completely impermeable to a solute $\pi' = \pi$ and $\sigma = 1$. If the membrane is completely permeable to a solute $\pi' = 0$, $\sigma = 0$; thus $1 \geqslant \sigma \geqslant 0$. The reflection coefficient is dimensionless. For biological membranes and small solutes σ is usually between 0.75 and 1.0. The osmotic pressure relationship, incorporating the reflection coefficient, can now be written

$$\pi = \sigma RT\Delta c_j \,. \qquad [2.2]$$

The value of the reflection coefficient has been employed for calculating the size of postulated water-filled pores in biological membranes (see p. 62).

Driving forces on ions

The discussion has so far dealt with the simple situation pertaining to uncharged solutes involved in passive transfer. If a substance is transported by diffusion alone, a linear relationship exists between the rate of transport and the concentration difference, any deviation from linearity indicating the action of another transport process (see p. 62). Although Fick's law is suitable for describing the passive movement of uncharged particles we are concerned here primarily with electrolytes and must therefore consider how these charged ions will behave.

An ion in aqueous solution is acted on by at least two physical forces, one arising from chemical potential gradients and the other from electrical potential gradients. Chemical potential is related to concentration as already discussed above, and electrical potential is the result of net positive or net negative charge carried by the ion. When a salt diffuses in water one of its constituent ion species generally has a higher mobility than the other and tends to diffuse faster than its oppositely charged partner. This situation results in a tendency towards charge separation, and thus a gradient of electrical potential, which causes a diffusion potential within the solution. Under the influence of this potential the faster moving ion of the pair is slowed down and the slower one is speeded up, and as a result they both move at the same rate. Thus a dissociated salt diffusing in a solvent behaves as a single substance, obeys Fick's law and has a characteristic diffusion coefficient.

In living systems the plasma membrane is usually the main barrier to free diffusion of particles into and out of cells. The passive movement of ions across such a membrane barrier results in a diffusion potential termed the *membrane diffusion potential*, the magnitude of which can be predicted in considering the driving forces for ion migration.

The driving force for an ion is the gradient of electrochemical potential and thus consists of two terms, one depending on the gradient of concentration and the other on the gradient of electrical potential. For an ion species j the electrochemical potential $\bar{\mu}_j$ is given by

$$\bar{\mu}_j = \mu_j^* + RT\ln\gamma_j c_j + z_j F\Psi \,, \qquad [2.3]$$

where μ_j^* is the chemical potential of the ion j in its standard state, R the gas constant, T the temperature in K, γ_j the activity coefficient, c_j the chemical concentration, z_j the valency (with sign), F the Faraday constant, Ψ the electric potential. $\gamma_j c_j$ is equal to the chemical activity a_j, the collective term $RT\ln a_j$ sometimes being used. If the ion is an ideal solution only the concentration term c_j need be employed and the collective term in Eq. [2.3] is then $RT\ln c_j$.

Passive flux equilibrium: the Nernst equation

The force on an ion is the negative of the electrochemical potential gradient, which consists of two terms, one depending on the activity (or concentration) gradient $RT\ln a_j^o/a_j^i$ and the other on the electrical potential gradient $z_jF(\Psi^i - \Psi^o)$, the superscripts i and o referring to inside and outside the cell. At equilibrium the two forces are equal and opposite, i.e. when $\bar{\mu}_j^o = \bar{\mu}_j^i$

$$z_jF(\Psi^i - \Psi^o) = RT\ln\left(\frac{a_j^o}{a_j^i}\right), \qquad [2.4]$$

and the electrical diffusion potential at equilibrium therefore becomes

$$\Psi^i - \Psi^o = \frac{RT}{z_j F} \ln\left(\frac{a_j^o}{a_j^i}\right). \qquad [2.5]$$

This is the *Nernst equation*, which describes the equilibrium state when the tendency for an ion to move down its chemical potential gradient in one direction is balanced by its tendency to move down its electrical potential gradient in the opposite direction.

Putting $\Psi^i - \Psi^o = E_{Nj}$, the *Nernst potential* for ion species j, inserting actual numerical values for R, F and replacing the natural logarithm by 2.303 log, where log is the common logarithm to the base 10, the Nernst equation becomes

$$E_{Nj} = \frac{58}{z_j} \log\left(\frac{a_j^o}{a_j^i}\right) \text{ mV at } 18^\circ \text{ C}, \qquad [2.6]$$

or

$$\frac{59.2}{z_j} \log\left(\frac{a_j^o}{a_j^i}\right) \text{ mV at } 25^\circ \text{ C},$$

or

$$\frac{62}{z_j} \log\left(\frac{a_j^o}{a_j^i}\right) \text{ mV at } 37^\circ \text{ C (body temperature)}.$$

For some calculations the ratio of the activity coefficients γ_j^o/γ_j^i may be assumed to equal one and the ratio a_j^o/a_j^i becomes c_j^o/c_j^i, the ratio of the concentrations. This assumption is justified when the ionic strengths on the two sides of a membrane are similar, but leads to errors when large concentration differences exist.

The Nernst potential E_{Nj} for an individual ionic species j may be calculated from the Nernst equation using the ratio of activities or concentrations. This value may be compared with the measured potential difference across a membrane E_M. If the measured and predicted (Nernst) values are the same then an equilibrium situation exists. If there is a marked difference between E_M and E_{Nj} then the system is not in equilibrium and energy must be expended to maintain the non-equilibrium state. The minimum amount of energy required is proportional to the difference between E_{Nj} and E_M in mV (1 mV equals 23.06 cal mol^{-1}). This difference ΔE_j, may be used to determine the direction of the net driving force. For cations a positive value for ΔE_j indicates an outward passive driving force, and any inward transport would need to do work against this, i.e. it would require a metabolically-driven active transport. Conversely, a negative ΔE_j would indicate a passive cation driving force inwards. The above situation is of

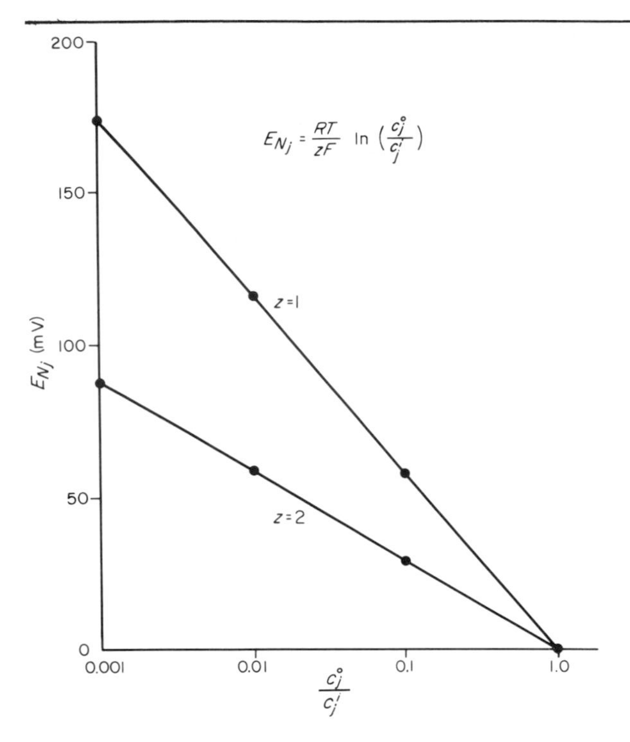

Fig. 2.2 Nernst potential E_{N_j} for monovalent ($z = 1$) and divalent ($z = 2$) ions, as a function of c_j^o/c_j^i, at 18°C. For cations, E_{N_j} is negative and for anions E_{N_j} is positive.

course reversed for anions, negative values of ΔE_j indicating an inward passive driving force and positive values indicating an outward one (see p. 107).

To illustrate further the Nernst principle let us consider a situation in which potassium ions are present at a concentration of 1 mol m^{-3} outside and 10 mol m^{-3} inside a membrane. From the Nernst equation it can be simply calculated that an electrical potential of –58 mV (negative inside) would maintain that concentration gradient across a membrane in a passive equilibrium state at 18° C. On the other hand, the maintenance of the same concentration gradient for chloride ions would require an electrical potential difference of +58 mV. This relationship between concentration and Nernst potential is depicted in Fig. 2.2, from which it can be seen that the Nernst potential increases by 58 mV for each 10-fold decrease in the concentration ratio. From the above simple example it will be apparent, first, that a relatively small electrical potential difference can balance a large concentration difference across a membrane, and secondly, that the existence of a concentration difference is not evidence that work has been performed in moving a particular ion species. The interior of most cells is electrically negative and, given an appropriate concentration gradient, cations will be able to enter passively, whereas anions will generally be moved in against the gradient of electrochemical potential and thus require metabolic energy for their transport, i.e. the transport will be active rather than passive (see p. 108).

Where the measured and predicted values of the Nernst potential differ, it is an indication that an ion is not distributing itself according to the Nernst equilibrium, which is not related to the properties of the membrane. It is possible that either the membrane is impermeable or not selectively permeable to that ion species or that some form of active transport is involved. If in the example given above for potassium the measured value was less negative than –58 mV, say –28 mV, then work would have to be done to pump potassium into the cell to maintain the internal concentration at 10 mol m^{-3}. In the absence of such a pump, potassium would leak out until the equilibrium concentration difference which could be maintained by –28 mV was reached. This value can be obtained from Fig. 2.2. If on the other hand the measured value was more negative, say –90 mV, then work would have to be done to prevent potassium from reaching a higher level than 10 mol m^{-3} inside, and the ion would have to be pumped out.

Flux ratio equation

A more critical way of identifying active transport is to determine the flux ratio for an ionic species across a membrane. This may be conveniently measured by using suitable radioactive isotopes and following the initial influx and efflux (see p. 44) to give the resultant net flux:

net flux = influx – efflux

or in symbols

$$\phi_j = \phi_j^{oi} - \phi_j^{io}.$$

Under steady-state conditions in certain non growing cells the net flux may be zero, influx and efflux being equal, but this situation may require work being performed to counter passive fluxes. The ratio of influx to efflux will be unity when the electrochemical potential across the membrane is zero, or differ from unity when a gradient of electrochemical potential exists. This relationship between flux ratios and driving forces on ions is given in an equation derived independently by Ussing (1949) and by Teorell (1949), known as the *Ussing–Teorell equation*:

$$\frac{\phi_j^{oi}}{\phi_j^{io}} = \frac{c_j^o}{c_j^i \exp(z_j F E_M / RT)} . \qquad [2.7]$$

By taking logarithms of both sides and substituting $\bar{\mu}_j$, the difference in electrochemical potential, we obtain

$$RT \ln \left(\frac{\phi_j^{oi}}{\phi_j^{io}} \right) = \bar{\mu}_j^o - \bar{\mu}_j^i . \qquad [2.8]$$

When $\bar{\mu}_j^o = \bar{\mu}_j^i$, $\phi_j = 0$ and the above equation reduces to the Nernst equation. Thus when E_M equals E_{Nj}, $\phi_j^{oi} = \phi_j^{io}$, no net passive flux of ion j will occur and no energy need be expended in moving the ion from one side of the membrane to the other. When the Ussing–Teorell equation is not satisfied, such ions are not moving passively or in some cases not moving independently from other fluxes. Interdependent fluxes can be described by irreversible thermodynamics which will be introduced later in this chapter (p. 52).

The use of the Nernst and Ussing–Teorell equations in indicating whether or not active transport of specific ions is taking place is illustrated in Chapter 5 (p. 106), where data obtained from experiments with large algal coenocytes are presented.

These criteria for active transport tell us nothing directly of the rate of transport of ions across the membrane, a process which is dependent on a number of factors, particularly the permeability of a membrane for a specific ion P_j, which is given by

$$P_j = \frac{u_j K_j RT}{\delta} , \qquad [2.9]$$

where μ_j is the mobility of the ion within the membrane, K_j the partition coefficient of the ion between water and the membrane phases, R and T as before and δ the membrane thickness. Although mobilities, solubilities and membrane thickness are difficult or impossible to measure in biological systems, P_j is relatively easy to estimate from tracer experiments involving passive ion fluxes. The permeability coefficient P_j, expressed in units of m s^{-1}, is in fact the sum of the permeabilities of a number of pathways which may be followed by an ion in its passage across a membrane and is therefore subject to some environmental modification (see p. 59).

Constant-field equation for membrane potential

Membranes normally separate a number of charged ions, the resultant diffusion potential which arises (see p. 69) being a reflection of both the distribution and permeability of the ions involved (see p. 59). If it is assumed that there is no net electric current flow through the membrane and that the potential gradient is uniform across the membrane, i.e. the electric field is constant, then considering only the major physiologically important ions, potassium, sodium and chloride, the membrane potential is given by:

$$E_M = \frac{RT}{F} \ln \left(\frac{P_K c_K^o + P_{Na} c_{Na}^o + P_{Cl} c_{Cl}^i}{P_K c_K^i + P_{Na} c_{Na}^i + P_{Cl} c_{Cl}^o} \right). \qquad [2.10]$$

This is the *Goldman* or *Hodgkin–Katz equation*, which is widely used in interpreting membrane potential differences. It is beyond the scope of the present work to give the derivation of this equation, the importance of which is that it predicts the value of the membrane potential as determined solely by passively moving ions.

Electrogenic transport across membranes

In virtually all biological membranes the active transport of some of the ion species involved may also contribute to the membrane potential if the active transport involves the transport of a charged ion-carrier complex on one or both of its journeys across the membrane. Such a process is termed *electrogenic* (see p. 70) and an additional term may be added to the membrane potential E_M so that

$$E_M = E_{eq} + E_x , \qquad [2.11]$$

where E_{eq} is the diffusion equilibrium potential and E_x an additive electrogenic mechanism. As in most biological membranes $P_K \approx P_{Cl} > P_{Na}$, the Goldman equation often gives a similar result to the Nernst equation for potassium and chloride. Because of the assumptions which underline the Goldman equation, differences in calculated and predicted membrane potentials do not necessarily indicate an electrogenic pump, and in most cases additional evidence is required before electrogenicity can unequivocally be claimed. If the blocking of an ion pumping mechanism with a metabolic inhibitor produced an abrupt change in the membrane potential, an electrogenic process may be inferred. Results obtained by Slayman (1970) for the fungus *Neurospora* illustrate this electrogenic phenomenon (Fig. 2.3). The membrane potential of *Neurospora* is rapidly decreased (depolarized) from about –220 mV to –40 mV when a proton extrusion process is inhibited by sodium azide, the potential returning to its original value when the inhibitor is removed and the pumping process unblocked.

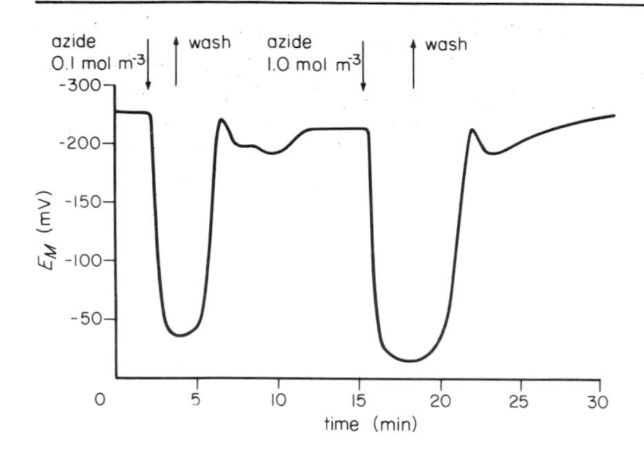

Fig. 2.3 The effect of sodium azide on the membrane potential E_M of the fungus *Neurospora.* An electrogenic proton extrusion pump is inhibited and the membrane is rapidly depolarized. (After Slayman, 1970.)

Evidence for the electrogenicity of the sodium extrusion pump in other systems is obtained by raising the internal sodium level of a cell and then observing the effect of placing the cell in a low sodium medium. The transfer of sodium should give rise to an increased membrane potential, and evidence that this is the case has been obtained by Thomas (1969), who observed a 20 mV increase of E_M when high sodium levels were introduced into snail neurones. This response could be blocked by ouabain treatment, and voltage clamp studies (see p. 44) indicated that some 28% of introduced sodium was extruded electrogenically.

Conductance of membranes

Information on the structure and organization of cell membranes and the manner in which ions cross them may be inferred from the electrical *conductance* g_M or its reciprocal *resistance*, as measured when a direct current is passed through the membrane, using the Ohm's law relationship ($g_M = dJ/dE_M$, dJ being the change of net current, dE_M the change in p.d.). Conductance may also be estimated from a knowledge of the independent ion fluxes across a membrane and the driving force. When a particular ion species is at electrochemical potential equilibrium, then a flux ϕ_j will contribute a partial ion conductance g_j such that

$$g_j = \frac{z_j^2 F^2}{RT} \varphi_j \, . \qquad [2.12]$$

For example a flux of 10 nmol $m^{-2} s^{-1}$ of potassium will contribute 38×10^{-3} mho m^{-2} partial conductance at 20° C. The resistance of the membrane R_j is the reciprocal of the conductance and would be 260×10^4 kΩm^2.

Estimates of membrane conductance or resistance obtained from the above relationship may be compared with direct electrical measurements obtained by passing a known current across the membrane. In squid axon, predicted values of g_j for potassium g_K, gave 0.2 mho m^{-2}, while the measured conductance was about 5 mho m^{-2}, 25 times greater (Hodgkin and Keynes, 1955). In giant algal cells g_K values of 40×10^{-3} mho m^{-2} have been calculated for the plasma membrane conductance, whereas the measured conductance is 500–1,000 $\times 10^{-3}$ mho m^{-2} (Findlay and Hope, 1964). These observations may be explained in two ways, either the potassium does not carry all the current or the ion movement is not independent of the movement of other ions. Considering the first possibility, the fluxes of sodium, chloride and calcium would not contribute more than a few additional mmho m^{-2}, but it has been suggested that proton fluxes may contribute a large part of the conductance. To investigate the possibility of interdependent movement, squid axon was treated with dinitrophenol (DNP), a respiratory uncoupler, to abolish any metabolically driven ion fluxes, and the g_K calculated as 6.6 mho m^{-2} against a measured value of 30 mho m^{-2} (Hodgkin and Keynes, 1955). The DNP probably inhibited an active sodium extrusion which was coupled to the potassium influx, the results therefore supporting the view that the movement of potassium in this system was not independent of other fluxes.

When current is passed through a membrane the membrane potential is changed from its resting value and is either increased (hyperpolarized) or decreased (depolarized) according to the direction of the current flow. Hyperpolarization cannot proceed above membrane potentials of about 300 mV, beyond which a phenomenon termed '*punch-through*' occurs, as illustrated in Fig. 2.4. Such a relationship is consistent with the view that the membrane is a *semiconductor*, composed of alternating layers of opposite fixed charge. Between these layers a depletion layer is postulated which increases in width as the membrane potential increases until it extends to the outer limits of the membrane, at which point 'punch-through' takes place and a sudden increase in current occurs. Such a system would permit rectification of about the observed amount to occur (see p. 52).

Methods of measurement

From the above discussion it will have become apparent that it is necessary to measure a limited number of parameters in order to estimate the driving forces and related electrical phenomena in living biological systems. With a

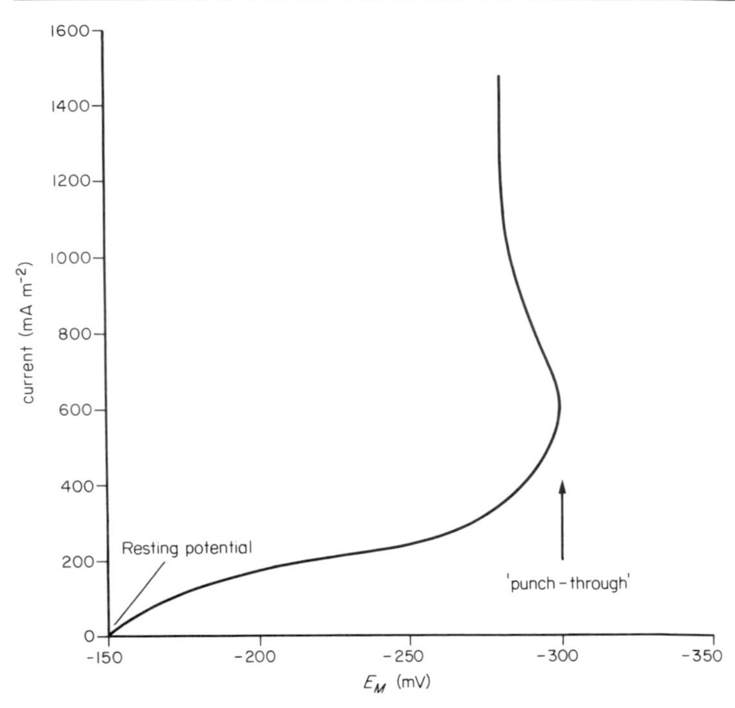

Fig. 2.4 The effect of an increased current on the membrane potential difference E_M. The resting potential was −150 mV. 'Punch-through' takes place when E_M = −300 mV.

knowledge of the ion concentration on either side of the membrane and the membrane electrical potential it is possible to calculate whether a Nernst-type passive equilibrium exists or if there is evidence of some form of metabolically-dependent active transport. In large cells, such as squid axon and algal coenocytes, it is possible to extract the cell contents and measure the ionic concentrations or activities, but in smaller cells other strategies must be employed. *Ion-sensitive electrodes* are increasingly used, although considerable technical difficulties occur with interfering ions, tip blockage and difficulties of insertion, thereby providing problems and frustrations for the operator. Another method is to estimate the internal concentration of an ion from a *flux analysis* (see p. 44), but often such estimates involve a number of untested assumptions and should be received cautiously.

Membrane electrical potentials are normally measured by inserting a glass microelectrode into a cell by micromanipulation under a microscope. The electrode is usually filled with 3×10^3 mol m^{-3} KCl and is coupled through a high input resistance electrometer (resistance 10^{10}–$10^{14}\Omega$) by

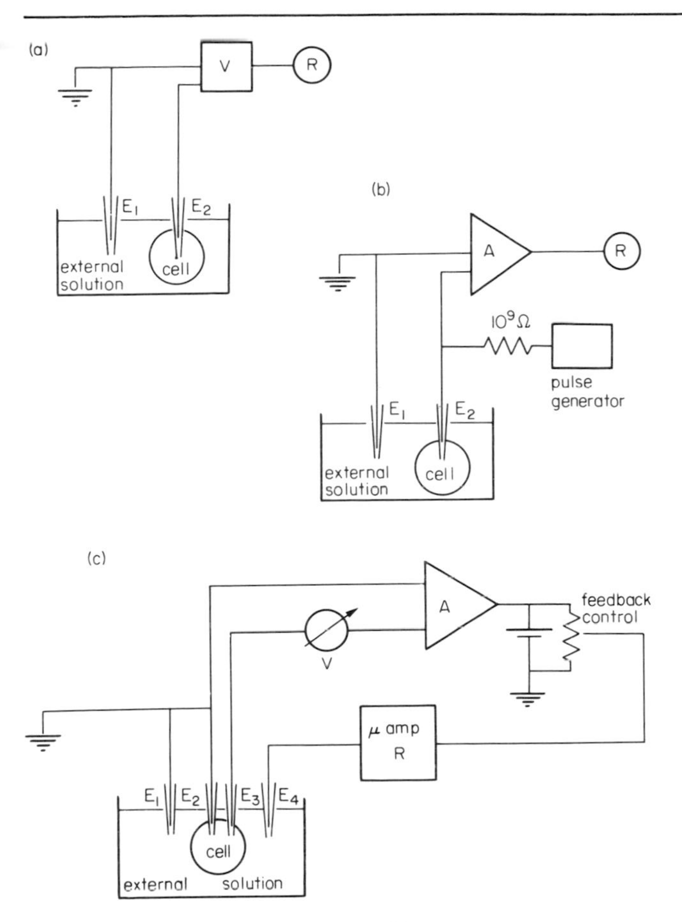

Fig. 2.5 Arrangement for measuring (*a*) electrical potential difference; (*b*) resistance; (*c*) the voltage-clamp or short-circuit technique. V is a high impedence voltmeter; R a recorder or oscilloscope; E_1 and E_4 are external or reference electrodes; E_2 and E_3 are internal microelectrodes; and A is a differential amplifier.

silver/silver chloride or calomel cells to a reference electrode placed in the external solution (Fig. 2.5*a*). The observed potential, normally negative inside with respect to outside, is of the order 30–200 mV for the majority of membrane systems on which measurements have been made.

Resistance measurement normally involves the insertion of two micro-electrodes into the cell, one to monitor the membrane potential and the other to pass the known current. To place two microelectrodes in a small cell is technically very difficult, and a method has been developed using a single electrode which utilises brief pulses of current of about 50 μs

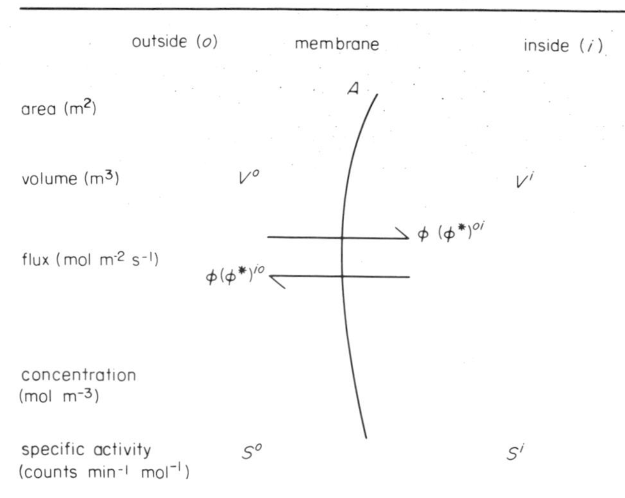

Fig. 2.6 The quantities considered in relating specific activity to time during the course of labelling or eluting a one-compartment system.

duration, permitting both potential and resistance to be measured (Fig. 2.5*b*).

Another electrical method which is frequently employed in ion transport studies is the *short-circuit technique* in which a voltage clamp device passes current across the membrane with zero chemical potential difference across it to maintain the potential difference at zero. Under such conditions any net ion movement will be active and can be measured by the magnitude of the short-circuit current (Fig. 2.5*c*).

Efflux analysis

The influx and efflux of ions across cell membranes may be measured using radioactive tracers and compared with the behaviour of model systems. The simplest model is a one-compartment system in which a cell of volume V, surface area A, containing a solution of solute j at a concentration c_j^i, is bounded by a membrane. The fluxes of j, or a radioactive isotope j^*, across the boundary membrane are depicted in Fig. 2.6. It is necessary to assume that the cell membrane does not distinguish between the radioactive and non-radioactive isotopes of the same element, so that tracer and stable ions enter the cell in the same ratio as occurs in the outside solution. The ratio of c_j^* to c_j is known as the specific activity S_j. Also the analysis assumes that rapid diffusion occurs on either side of the surface boundary, and thus the gradient of solute j is across the membrane only. Under these conditions the rate of accumulation of tracer is given by

$$\frac{V\mathrm{d}c_j^{i*}}{\mathrm{d}t} = A\left[\phi_j^{oi}\frac{c_j^{o*}}{c_j^o} - \phi_j^{io}\frac{c_j^{i*}}{c_j^i}\right]. \qquad [2.13]$$

Influx is determined by measuring the initial rate of increase of internal radioactivity when an unlabelled cell is placed in a radioactive solution. As c_j^{i*} is much smaller than c_j^i, the tracer efflux is negligible and Eq. [2.13] can be rewritten for influx to give

$$\phi_j^{oi} = V\left(\frac{\mathrm{d}c_j^{i*}}{\mathrm{d}t}\right)\Big/A\left(\frac{c_j^{o*}}{c_j^o}\right). \qquad [2.14]$$

Efflux is determined by allowing the cell to accumulate tracer from a labelled bathing solution, briefly rinsing and then placing it in an identical non-radioactive bathing solution, which is replaced frequently so that c_j^{o*} remains zero. In this situation the loss of radioactivity from the cell is exponential, as the internal specific activity is decreasing during the efflux period, and thus

$$\ln(Vc_j^{i*}) = \frac{A\phi_j^{io}}{Vc_j^i}\,t. \qquad [2.15]$$

When $\ln(Vc_j^{i*})$ is plotted against time a straight line should result, the slope of which gives the efflux value.

In practice, the efflux is often non-linear (Fig. 2.7), and extrapolation of the linear portion of efflux curves to the ordinate may give an indication of the amount of tracer in different phases or compartments of the cell. The slope of those linear portions gives the efflux across membranes in series, as for instance occurs in the cells of plants where the inner vacuolar compartment is bounded by the tonoplast. This efflux method for compartmental analysis in plant cells is dependent on the relative permeabilities of the plasma membrane and tonoplast. If the plasma membrane is more permeable than the tonoplast, then during an influx period the cytoplasm will fill up with isotope first and then afterwards the vacuole will fill, enabling the separate influxes to be measured. On efflux the phases will be represented by the linear portions of the efflux curve presented in Fig. 2.7. However, the situation becomes more complex to analyse if the tonoplast is much more permeable than the plasma membrane, and under such conditions the plasma membrane becomes the rate-limiting barrier for both influx and efflux determinations.

Kinetics of ion transport

It has been observed that the kinetics of ion transport across membranes in a wide range of cells are similar to those of enzymic catalysis, and this

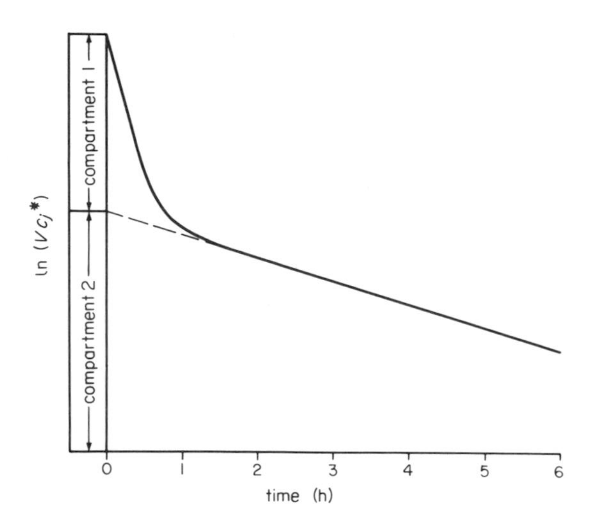

Fig. 2.7 The total radioactivity Vc_j^{i*} plotted against time for the efflux from a two-compartment system. The extrapolation of the linear portion to zero time gives the quantity of tracer in the two compartments. The efflux from each compartment may be determined from the slopes of the linear portions of the curve.

has led to the proposal that reversible binding to a carrier mediates the transport process. In both ion transport and enzyme catalysis the mechanism is believed to involve the attachment of a substrate (ion or enzyme substrate) to an active agent (ion carrier or enzyme) which is released following transport or catalysis to recombine with the substrate once again.

The rate of *carrier-mediated ion transport* can therefore be characterized by two factors, one the maximum rate of transport which can be achieved when all the available carrier sites are loaded V_{max}, and the other the fraction of the carrier actually loaded at a given substrate concentration θ. V_{max} can be calculated from the asymptote reached when the rate of absorption is measured over a range of concentrations (Fig. 2.8) and θ, the fraction of sites occupied at a given ion concentration [S], is obtained from the Langmuir adsorption equation:

$$\theta = \frac{[S]}{K_m + [S]}, \qquad [2.16]$$

where K_m is the dissociation constant of the carrier–ion complex, characteristic of a particular ion crossing a specific membrane and is expressed in units of concentration (mol m^{-3}). The rate of absorption ν is given by the product of the two factors, V_{max} and θ, to give

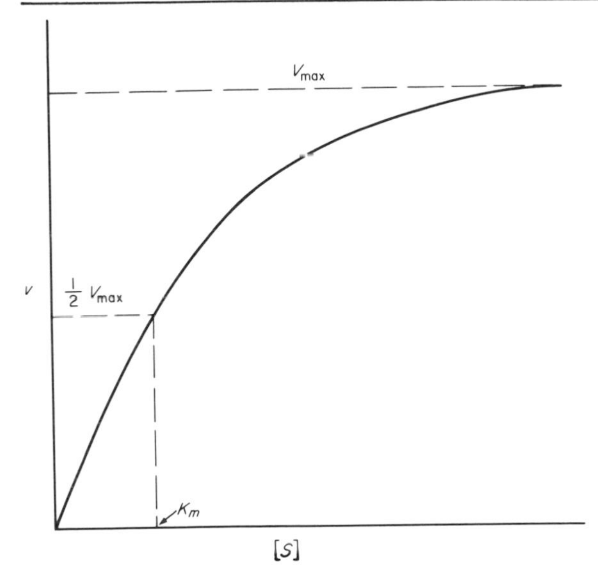

Fig. 2.8 The relationship between the external solute concentration [S] and the rate of absorption v.

$$v = \frac{V_{max}[S]}{K_m + [S]}. \qquad [2.17]$$

This relationship is based on the assumption that the model for uptake is as follows:

$$S + C \underset{k_{-1}}{\overset{k_1}{\rightleftharpoons}} SC \underset{k_{-2}}{\overset{k_2}{\rightleftharpoons}} C + S \qquad [2.18]$$

where S represents the ion, C the carrier and k_1, k_{-1}, k_2, and k_{-2} the rate constants of the reaction. Eq. [2.18] is dependent on there being little or no counterflow from right to left, k_{-2} being negligible.

By analogy with enzyme kinetics, K_m is equal to the concentration of the ion [S] at which V reaches half the theoretical maximal rate. Substituting K_m for [S] in Eq. [2.17], we have.

$$v = \frac{K_m V_{max}}{2K_m} = \tfrac{1}{2} V_{max}. \qquad [2.19]$$

Eq. [2.17] is known as the *Michaelis–Menten equation* and can be rearranged to yield two straight-line plots, the double reciprocal form due to Lineweaver and Burk (1934)

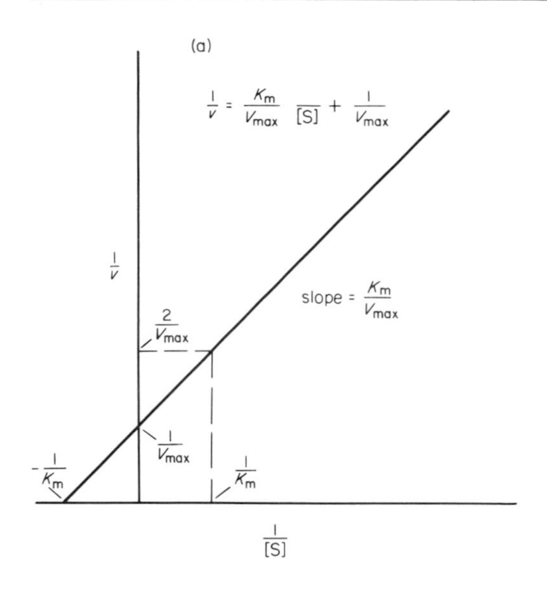

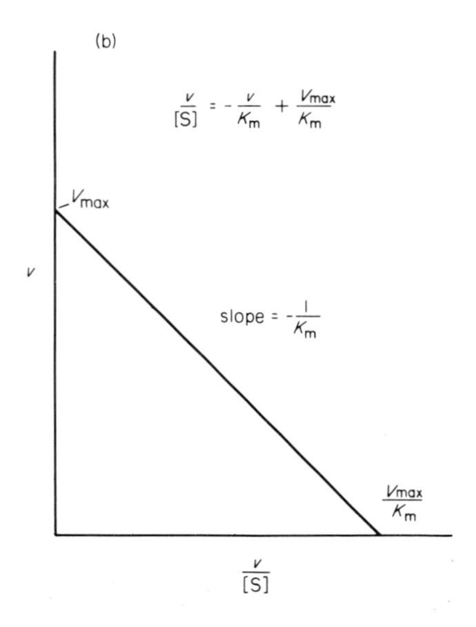

Fig. 2.9 Linear transformations of the Michaelis–Menten equation. (*a*) Lineweaver and Burk (1934); (*b*) Eadie, Hofstee (1952).

$$\frac{1}{\nu} = \frac{K_m}{V_{max}[S]} + \frac{1}{V_{max}}, \qquad [2.20]$$

and the single reciprocal form of Eadie and of Hofstee (1952)

$$\frac{\nu}{[S]} = \frac{V_{max}}{K_m} - \frac{\nu}{K_m}$$

or

$$\nu = V_{max} - \frac{K_m}{\nu[S]}; \qquad [2.21]$$

also

$$\frac{[S]}{\nu} = \frac{K_m}{V_{max}} + \frac{[S]}{V_{max}}.$$

The way in which these equations may be employed for the determination of V_{max} and K_m is indicated in Fig. 2.9. Plotting of data by Eq. [2.20] gives $1/\nu$ versus $1/[S]$, which appears to have some advantages, but determination of values for V_{max} and K_m requires statistical treatment. Data can be fitted to Eq. [2.21], plotting ν versus $\nu/[S]$, by the least-squares method using ν^4 weighting factors.

The constant K_m has been called a constant of convenience in that it may not be a true rate, affinity or dissociation constant, merely an operational term. However, it enables investigators to characterize the affinity of a 'carrier' for a substrate. A low K_m value indicates a high affinity, while a high K_m indicates a low affinity. If two substances have an affinity for the same carrier site, they will inhibit each other *competitively.* With this type of inhibition the K_m for the uptake of the substances is altered, the V_{max} however remaining the same (Fig. 2.10*a*). If the transport of a substance is inhibited by a metabolic inhibitor such as 2,4-DNP there is no effect on the K_m but V_{max} is reduced. Such an inhibition is termed *non-competitive* (Fig. 2.10*b*).

If a substance were transported by more than one carrier site the uptake isotherm would not be a single hyperbola, as in Fig. 2.8, but might show two or more hyperbolae resulting in a similar number of K_m and V_{max} values. Such multiple-carrier processes, often limited to only two, have been indicated in many animal and plant cells, and in bacteria.

In a number of cases data have been obtained which do not conform to the Michaelis–Menten equation. In the erythrocyte, uptake of K^+ may show a sigmoidal rather than a hyperbolic relationship (Garrahan and Glynn, 1967). Such data have been interpreted in terms of biochemical allosteric interactions and Hill plots have been used to determine the cooperativity of the interaction (Fig. 2.11; Koshland, 1970). The Hill relationship is

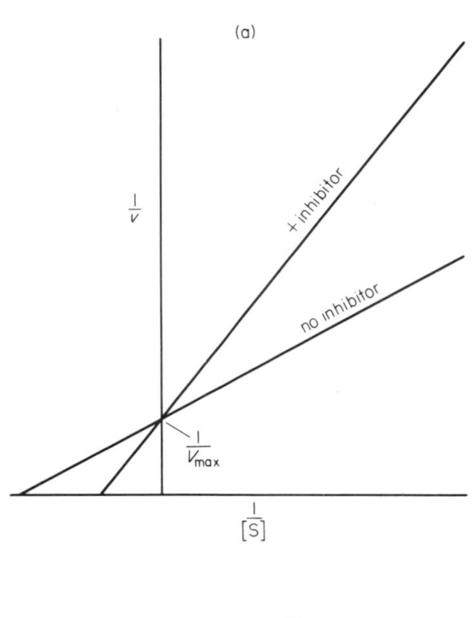

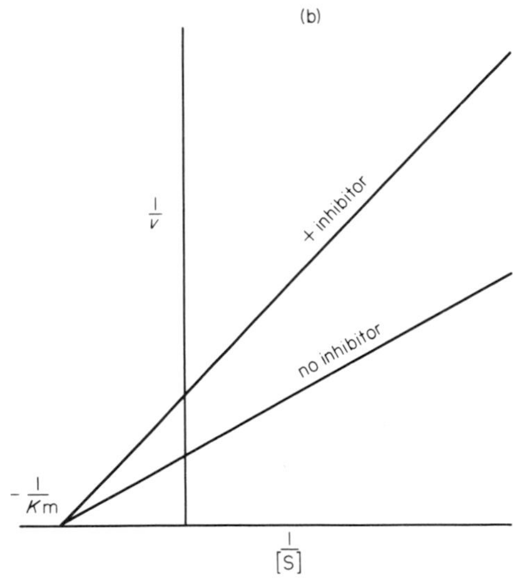

Fig. 2.10 Lineweaver and Burk plots of (*a*) competitive inhibition; (*b*) non-competitive inhibition.

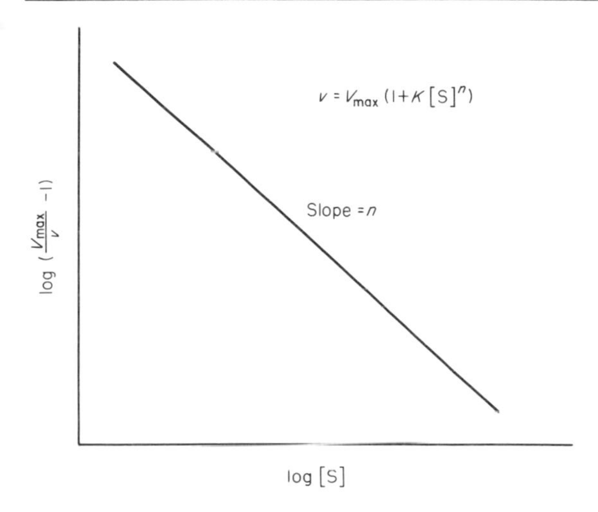

Fig. 2.11 Hill plot for the determination of the co-operativity of the interaction between substrate and enzyme.

$$v = V_{max}\,(1 + K[S]^n)\,,$$

where [S] is the activator concentration, n the interaction coefficient, K the apparent dissociation equilibrium constant and v and V_{max} are as previously defined. $n > 1$ indicates positive co-operativity, while $n < 1$ indicates negative co-operativity and $n = 1$ is equivalent to the Michaelis–Menten equation.

Such comparisons indicate that ion absorption isotherms which are the result of active processes need not follow the Michaelis–Menten formalism, nor can too much significance be attached to the observation that a particular isotherm does so. Passive fluxes, particularly if carrier mediated, may under certain conditions manifest saturation kinetics and therefore produce a rectangular hyperbola and apparent Michaelis–Menten kinetics (p. 62).

An interesting alternative approach to ion uptake kinetics has been developed by Thellier (1970), who has analysed absorption kinetics, including some of the dual carrier processes referred to above, without invoking the concept of carrier sites. The overall uptake is envisaged as

$$S^o \underset{}{\overset{\text{cell}}{\rightleftharpoons}} S^i\,,$$

where S^o and S^i represent the substrate (ion) outside and inside the cell. The velocity of the process v is formally equivalent to an electric intensity I, and the magnitude is given by:

$$\Delta E = 2.3\, A \log \left[B \frac{[S^o]}{[S^i]} \right], \qquad [2.22]$$

where ΔE is the formal equivalent of the electrical potential difference, A equals RT/zF when electric charges are transferred and B is a constant characteristic of the thermodynamic state of the cell.

With a process which obeys Ohm's law

$$I = \Delta E / r\,, \qquad [2.23]$$

where r is the resistance, and then

$$\nu = 2.3 \frac{A}{r} \log \left[B \frac{[S^o]}{[S^i]} \right]. \qquad [2.24]$$

When the process is non-ohmic as with varistant semiconductors or with tissue when $[S^o]$ is high,

$$I = \Delta E / r + (\lambda \Delta E)^m\,, \text{ where } m > 1 \qquad [2.25]$$

and

$$\nu = 2.3 \frac{A}{r} \log \left[B \frac{[S^o]}{[S^i]} \right] + 2.3\, \lambda\, A \log \left[B \frac{[S^o]^m}{[S^i]} \right], \qquad [2.26]$$

where λ and m are parameters characteristic of cell structures catalyzing the processes.

When the above equations are applied to data on ion absorption showing dual isotherms, an ohmic process is indicated when $[S^o]$ is low and a non-ohmic process when $[S^o]$ is high. This analysis would indicate that dual isotherms are the result of structural changes within the membrane, which is seen to behave as a semiconductor (see p. 41) thus making it unnecessary to invoke the presence of two or more carrier systems.

Irreversible thermodynamics of transport processes

Earlier in this chapter (p. 37) it was pointed out that when ions are not moving independently from other fluxes the Ussing–Teorell equation, [2.7], may not be satisfied. Under such conditions transport may still be passive, but due to *interdependent fluxes.*

Water, ions and other solutes moving through a membrane exert a frictional drag on each other and their fluxes therefore are interdependent. That is to say that the flux of a solute will be dependent not only on its own gradient of chemical potential but also on the gradient of the chemical potential of the solvent water. These interdependent fluxes can be quantitatively described by irreversible thermodynamics. Classical thermo dynamic treatment is restricted to reversible processes, or systems at equilibrium, and can rarely be applied to living biological systems, which exhibit few equilibria.

The general theory of irreversible thermodynamics concerns the relationships between flows or fluxes of matter, or energy, or electrical charge, within a system and the forces responsible for them. Onsager (1931) proposed that the flux of a component is linearly dependent on all the thermodynamic forces operating in the system, although this linearity holds only for processes not going too fast, or not too far from equilibrium. A set of equations can be written relating all the fluxes and forces in the following form:

$$\begin{aligned} \phi_1 &= L_{11}X_1 + L_{12}X_2 + \ldots + L_{1n}X_n \\ \phi_2 &= L_{21}X_1 + L_{22}X_2 + \ldots + L_{2n}X_n \\ &\vdots \\ \phi_n &= L_{n1}X_1 + L_{n2}X_2 + \ldots + L_{nn}X_n \end{aligned} \qquad [2.27]$$

where each flux ϕ_i is dependent on its conjugate force X_i through a straight coefficient L_{ii} which is always positive, and may also be linked to non-conjugate forces X_j through the cross-coefficients L_{ij} $(i \neq j)$, which may be positive or negative as long as the determinant formed by the co-efficients is positive. An important reduction in the number of coefficients comes about because of the *Onsager reciprocal relation*, which states that the 'matrix' of the cross-coefficients is symmetrical so that $L_{ij} = L_{ji}$, corresponding to having equal action and reaction. The reciprocal relation requires that the rate of entropy production multiplied by temperature, per unit volume, Φ is given by the sum of the products of these fluxes and forces, that is

$$\Phi = \sum_j X_j \phi_j \cdot \qquad [2.28]$$

Although Onsager had established a sound and convenient means of coupling linked flows with the forces responsible for them, it was not until the work of Staverman (1952) that this approach was applied to the transport of materials across membranes. To illustrate the principles involved we will first consider the simple case of a single non-electrolyte solute moving across a membrane barrier, and then we will consider the more complex situation which arises when the system contains electrolytes. In order to keep the analysis in a manageable form we will restrict the treatment to isothermal conditions.

If a membrane is permeable to both water (w) and solute (s), the fluxes can be represented by two phenomenological equations:

$$\begin{aligned} \phi_w &= L_{ww}X_w + L_{ws}X_s \\ \phi_s &= L_{sw}X_w + L_{ss}X_s \,. \end{aligned} \qquad [2.29]$$

The forces X_w and X_s can be replaced by the 'reduced forces' $\Delta\mu_w$ and $\Delta\mu_s$ (Kedem and Katchalsky, 1963), where $\Delta\mu_w$ and $\Delta\mu_s$ are the differences in the chemical potentials of water and solute, respectively.

$$\Delta\mu_w = \overline{V}_w(\Delta P - \Delta\Pi_s) \qquad [2.30]$$

where $\overline{V}_w$ is the partial molal volume of water, ΔP is the difference in hydrostatic pressure and $\Delta\Pi_s$ is the difference in osmotic pressure corresponding to the difference in concentration Δc_s.

$$\Delta\mu_s = \overline{V}_s\Delta P + \Delta\Pi_s/\bar{c}_s \, , \qquad [2.31]$$

where $\bar{c}_s$ is an average concentration within the membrane defined by

$$\bar{c}_s = \Delta\Pi_s/RT\Delta\ln a_s \, . \qquad [2.32]$$

Thus

$$\phi_w = L_{ww}\Delta\mu_w + L_{ws}\Delta\mu_s$$

$$\phi_s = L_{sw}\Delta\mu_w + L_{ss}\Delta\mu_s \, . \qquad [2.33]$$

These equations can be transferred in terms of the *entropy dissipation function* Φ to give

$$\Phi = \phi_w\Delta\mu_w + \phi_s\Delta\mu_s$$

$$\doteqdot \phi_w\overline{V}_w\,(\Delta P - \Delta\Pi_s) + \phi_s\overline{V}_s\Delta P + \Delta\Pi_s/\bar{c}_s \, . \qquad [2.34]$$

When the membrane is permeable to the solute, ϕ_s is finite and volume change on either side of the membrane is due to both ϕ_w and ϕ_s. A volume flow ϕ_v is then observed

$$\phi_v = \phi_w\overline{V}_w + \phi_s\overline{V}_s \qquad [2.35]$$

Equation [2.29] can now be rearranged to give

$$\Phi = \phi_v\,(\Delta P - \Delta\Pi_s) + \phi_s\,(\Delta\Pi_s/\bar{c}_s) \, , \qquad [2.36]$$

and the new conjugate flows and forces can be written

$$\phi_v = L_{ww}\,(\Delta P - \Delta\Pi_s) + L_{ws}\,(\Delta\Pi_s/\bar{c}_s)$$

$$\phi_s = L_{sw}\,(\Delta P - \Delta\Pi_s) + L_{ss}\,(\Delta\Pi_s/\bar{c}_s) \qquad [2.37]$$

where L_{ww} and L_{ss} are the 'straight' coefficients linking volume and solute flow to ΔP and $\Delta\Pi_s$, respectively, and L_{ws} and L_{sw} are the 'cross' coefficients which describe the volume flow associated with $\Delta\Pi_s$ and the exchange flow associated with ΔP, respectively. It should be remembered that $L_{ws} = L_{sw}$ as before.

When $\Delta\Pi_s$ is zero, i.e. when there is no solute concentration across the membrane, $\phi_v = L_{ww}\Delta P$ and the observed flow is the result of ΔP only. Thus L_{ww} can be identified as the *filtration coefficient* (or the hydraulic conductivity L_p, units m s^{-1} bar^{-1}).

When $\Delta P - \Delta\Pi_s = 0$, i.e. when pressure is applied to stop volume flow, $\phi_s = L_{ss}\ (\Delta\Pi_s/\bar{c}_s)$ and L_{ss} can be identified as the *solute permeability coefficient* ω_s, units mol s^{-1} m^{-2}.

These two coefficients, L_p and ω_s, are only sufficient if the membrane is completely permeable or completely impermeable to the solute molecules. For the range of intermediate situations encountered in biological systems a selectivity or *reflection coefficient* σ is introduced (see p. 33). When $\phi_v = 0$,

$$L_{ww}\ (\Delta P - \Delta\Pi_s) = -L_{ws}\ (\Delta\Pi_s/\bar{c}_s)$$

and

$$\sigma = \frac{-L_{ws}}{L_{ww}} = \frac{\Delta P - \Delta\Pi_s}{\Delta\Pi_s/\bar{c}_s}\ . \qquad [2.38]$$

Equation [2.37] can now be written

$$\phi_v = L_p\ (\Delta P - \Delta\Pi_s) - \sigma L_p \Delta\Pi_s/\bar{c}_s\ . \qquad [2.39]$$

When $L_{ww} = L_{ws}$, $\sigma = 1$ and the membrane is impermeable to the solute. When the membrane is non-selective, $\sigma = 0$. For intermediate states $1 \geqslant \sigma \geqslant 0$. The reflection coefficient is dimensionless. Using the coefficients derived above the solute flow ϕ_s becomes

$$\phi_s = (1 - \sigma)\ \bar{c}_s\phi_v + \omega_s\Delta\Pi_s/\bar{c}_s\ . \qquad [2.40]$$

Thus the three independent coefficients required to provide a correct description of the above system are defined as:

$$L_p = \left(\frac{\phi_v}{\Delta P}\right)_{\Delta\Pi_s = 0}\ ,$$

$$\omega = \left(\frac{\phi_s}{\Delta\Pi_s/\bar{c}_s}\right)_{\Delta P - \Delta\Pi_s = 0}$$

$$\sigma = \left(\frac{\Delta P - \Delta\Pi_s}{\Delta\Pi_s/\bar{c}_s}\right)_{\phi_v = 0}\ .$$

When the solute under consideration is an electrolyte, in addition to the forces considered above the electrical force must also be included. The fluxes can then be represented by three phenomenological equations and the entropy dissipation function becomes

$$\Phi = \phi_w\Delta\mu_w + \phi_s\Delta\mu_s + J\Delta V\ , \qquad [2,41]$$

where J is the flow of electrical current through the membrane (A m^{-2}) and V is the electromotive force (volts).

The phenomenological equations for an electrolyte are

$$\phi_v = L_{ww}\ (\Delta P - \Delta\Pi_s) + L_{ws}\ (\Delta\Pi_s/\bar{c}_s) + L_{wv}\Delta V$$

$$\phi_s = L_{sw}\ (\Delta P - \Delta\Pi_s) + L_{ss}\ (\Delta\Pi_s/\bar{c}_s) + L_{sv}\Delta V$$
$$J = L_{vw}\ (\Delta P - \Delta\Pi_s) + L_{vs}\ (\Delta\Pi_s/\bar{c}_s) + L_{vv}\Delta V\ . \qquad [2.42]$$

For these three equations six independent Onsager coefficients are required. In addition to L_p, ω and σ, three electrical coefficients are required. When $\Delta P = 0$ and $\Delta\Pi_s = 0$ the coefficients are:

$L_{wv} = \phi_v/\Delta V$, the *electro-osmotic volume flow*;

$L_{sv} = \phi_s/\Delta V$, the *electro-osmotic solute flow*;

$L_{vv} = J/\Delta V = g$, the *specific electrical conductance* (mho m^{-2}).

When $\Delta\Pi_s = \Delta V = 0$,

$L_{ww} = J/\Delta P$, the *streaming current* per unit pressure difference,

and when $\Delta P - \Delta\Pi_s = 0$ and $\Delta V = 0$,

$L_{vs} = J/(\Delta\Pi_s/\bar{c}_s)$, the *permeation current.*

Other well-known parameters are obtained by certain combinations of the above coefficients. When $\phi_v = 0$ and $\Delta\Pi_s = 0$,

$\dfrac{-L_{wv}}{L_{ww}} = \dfrac{\Delta P}{\Delta V} = P_v$, the *electro-osmotic pressure* at zero flow per unit potential.

When $\Delta\Pi_s = 0$ and $\Delta P = 0$,

$\dfrac{L_{wv}}{L_{vv}} = \phi_v/J = \beta$, the *electro-osmotic permeability*;

and

$$\frac{L_{sv}}{L_{vv}} = \phi_s/J = \frac{\nu}{izF},$$

where ν is the transport number, z the valency, i the number of ions per salt molecule and F the Faraday.

With the above coefficients and combinations of coefficient, Katchalsky and Kedem (1962) have written the three phenomenological equations for an electrolyte as

$$\phi_v = L_p\ (\Delta P - \Delta\Pi_s) - \sigma L_p \Delta\Pi_s/\bar{c}_s - P_v L_p J/g$$
$$\phi_s = (1 - \sigma)\ \bar{c}_s \phi_v + \omega_s \Delta\Pi_s/\bar{c}_s + \frac{\nu J}{izF}$$
$$J = -P_v\phi_v + \frac{g\nu}{izF}\frac{\Delta\Pi_s}{c_s} + g\Delta V \qquad [2.43]$$

for the flow of volume, solute and electricity, respectively. When there is no flow of electricity $J = 0$ and Eqs. [2.43] reduce to Eqs. [2.39] and [2.40].

Active transport

Early workers in the field of ion transport believed that the finding of a higher concentration of ions inside a biological membrane as compared with that outside, in the medium, was evidence of an active accumulation of those ions. However, the inadequacy of this approach and its lack of specificity became apparent in that it gave no indication of the particular ion on which work was done, necessitating its replacement by a more rigorous definition. Ussing (1949) defined active transport as the process by which an ion is moved against a gradient of electrochemical potential, movement of the ion therefore being dependent on the decrease in free energy of a metabolic process. When the membrane conditions are known in sufficient detail the Ussing–Teorell flux ratio equation, [2.7], can be used to predict whether the net flux of an ion is passive or active. As the magnitude, but not the sign, of the logarithm of the flux ratio may be changed by non-independence and the interaction of ions in the membrane, active transport is identified by a flux ratio whose logarithm has the wrong sign, provided there is no obvious flow component to which the transport could be coupled.

Active transport has been defined in terms of irreversible thermodynamics as an entrainment between a transport flux and a metabolic reaction. This requires the inclusion of a reaction flux ϕ_r, driven by its affinity A_r, in the phenomenological equations for the system, active transport being characterized by a non-zero coupling coefficient R_{jr} between the flow of substance j and the reaction. It is very difficult to apply correctly the necessary thermodynamic conditions and these difficulties limit the practicability of the above method, the usual practical test for active transport remaining that of Ussing as defined above.

The biophysical approach to the problems of ion transport, as outlined in this chapter, has contributed a great deal to the study of ion transport but a drawback is that it has only been used to explain the transport properties of the living cell in terms of diffusion barriers and fixed compartments. Investigators have now characterized many ion pumps in these biophysical terms and current research is moving towards providing a conceptual understanding of ion transport processes in terms of the dynamic nature of the cell and its membranes.

Further reading and references

Further reading

BAKER, D. A. and HALL, J. L. (1975) Ion transport – introduction and general principles, in *Transport in Plant Cells and Tissues,* p. 1, eds D. A. Baker and J. L. Hall. North Holland, Amsterdam and London.

HOPE, A. B. (1971) *Ion Transport and Membranes – A Biophysical Outline.* Butterworths, London.

NEAME, K. D. and RICHARDS, T. G. (1972) *Elementary Kinetics of Membrane Carrier Transport.* Blackwell, Oxford.

NOBEL, P. S. (1974) *Introduction to Biophysical Plant Physiology.* Freeman, San Francisco.

SLAYMAN, C. L. (1970) Movement of ions and electrogenesis in micro-organisms, *Am. Zoologist,* **10,** 377.

STEIN, W. D. (1967) *The Movement of Molecules Across Cell Membranes.* Academic Press, London and New York.

USSING, H. H. (1969) *Active Transport in Theoretical Physics and Biology.* North Holland, Amsterdam and London.

Other references

EADIE, G. S. (1952) On the evaluation of the constants V_m and K_m in enzyme kinetics, *Science,* **116,** 688.

FINDLAY, G. P. and HOPE, A. B. (1964) Ionic relations of cells of *Chara australis.* VII. The separate electrical characteristics of the plasmalemma and tonoplast, *Aust. J. biol. Sci.,* **17,** 66.

GARRAHAN, P. J. and GLYNN, I. M. (1967) The sensitivity of the sodium pump to external sodium, *J. Physiol.,* **192,** 175.

HODGKIN, A. L. and KEYNES, R. D. (1955) Active transport of cations in giant axons from *Sepia* and *Loligo, J. Physiol.,* **128,** 28.

HOFSTEE, B. H. J. (1952) On the evaluation of the constants V_m and K_m in enzyme reactions, *Science,* **116,** 329.

KATCHALSKY, A. and KEDEM, O. (1962) Thermodynamics of flow processes in biological systems, *Biophys. J.,* **2,** 53.

KEDEM, O. and KATCHALSKY, A. (1963) Permeability of composite membranes. I. Electric current, volume flow and flow of solute through membranes, *Trans. Farad. Soc.,* **59,** 1918.

KOSHLAND, D. E. (1970) The molecular basis for enzyme regulation, in *The Enzymes,* 3rd edn, Vol. 1, p. 341, ed. P. D. Boyer. Academic Press, London and New York.

LINEWEAVER, H. and BURK, D. (1934) The determination of enzyme dissociation constants, *J. Amer. Chem. Soc.,* **56,** 658.

ONSAGER, L. (1931) Reciprocal relations in irreversible processes, *Physiol. Rev.,* **38,** 2265.

STAVERMAN, A. J. (1952) Non-equilibrium thermodynamics of membrane processes, *Trans. Farad. Soc.,* **48,** 176.

TEORELL, T. (1949) Membrane electrophoresis in relation to bioelectrical polarization effects, *Arch. Sci. Physiol.,* **3,** 205.

THELLIER, M. (1970) An electrokinetic interpretation of the functioning of biological systems and its application to the study of mineral salts absorption, *Ann. Bot.,* **34,** 983.

THOMAS, R. C. (1969) Membrane current and intracellular sodium changes in a snail neurone during extrusion of injected sodium, *J. Physiol.,* **201,** 495.

USSING, H. H. (1949) The distinction by means of tracers between active transport and diffusion, *Acta physiol. Scand.,* **19,** 43.

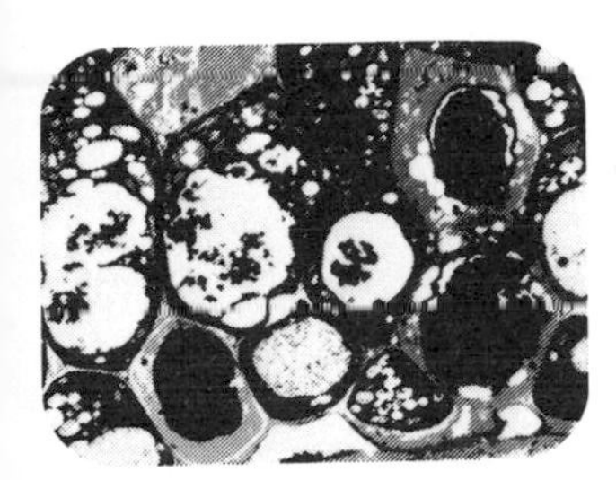

Chapter 3

Passive movements and ion selectivity

Permeability

In the preceding chapter the movement of ions across membranes was described in biophysical terms as being active or passive. Such passive movement is down the gradient of electrochemical potential and such a movement is diffusional. However the rate of this movement is dependent on the structure and composition of the membrane (see Ch. 1) and therefore its *permeability* to the diffusing molecule. The term permeability is thus a property of the membrane and is a measure of the ease with which a solute moves passively through a particular membrane. Diffusing molecules obey Fick's law (Eq. [2.1]), but as well as D_j, the diffusion coefficient, the concentration of j within the membrane must also be considered. This second factor is expressed as the *partition coefficient* K_j, which is the relative solubility of j within the membrane. The passive flux of a solute across a membrane can therefore be written as

$$\phi_j = D_j K_j \frac{(c_j^o - c_j^i)}{\delta}, \qquad [3.1]$$

where D_j is the diffusion coefficient of species j, K_j the partition coefficient of species j between the membrane phase and the bathing solvent water, and δ is the thickness of the membrane. As K_j and δ are not measurable with any great accuracy for biological membranes, they are combined with D_j to give a permeability coefficient P_j, which may be defined as

$$P_j = \frac{D_j K_j}{\delta}. \qquad [3.2]$$

The value of K_j may be determined by measuring the ratio of the equilibrium concentration of solute j in a lipid phase and an aqueous phase, this being based on the high lipid content of membranes (see p. 7).

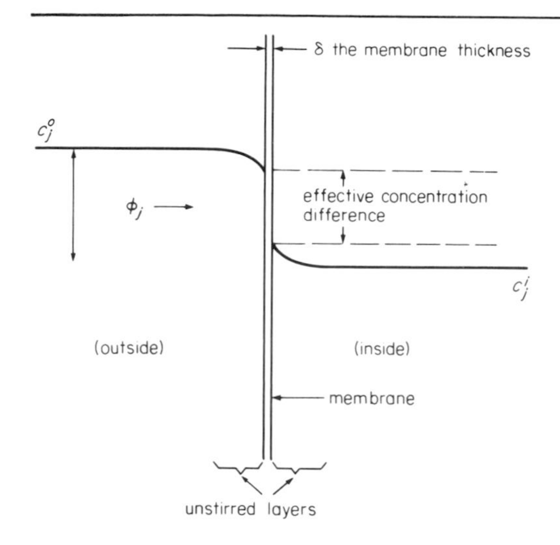

Fig. 3.1 The effect of unstirred boundary layers at the inner and outer surfaces of a membrane on the concentration of solute species j near a membrane across which the solute is diffusing.

Partition coefficients have a wide range of values, most being between 10^{-6} and 10. Small inorganic ions have low K_j values, close to 10^{-5}, whereas some non-polar organic solutes have values approaching unity. For this reason membranes are usually much less permeable to electrolytes than to non-electrolytes.

A working definition of P_j therefore becomes the number of molecules crossing unit area of a membrane in unit time when a concentration difference is applied across that membrane. Thus the net flux inwards ϕ_j (units mol m^{-2} s^{-1}) is related to P_j by combining Eqs. [3.1] and [3.2] to give

$$\phi_j = P_j\,(c_j^o - c_j^i) \cdot \qquad [3.3]$$

where c_j^o and c_j^i are the outside and inside concentrations of solute j, respectively. It can be seen from the above relationship that P_j has units of m s^{-1}.

A major problem associated with the practical measurement of P_j is that of determining the magnitude of the 'effective' concentration difference across a membrane. This determination is made somewhat uncertain by the existence of *unstirred boundary layers* at the inner and outer surfaces of a membrane, as illustrated in Fig. 3.1. Diffusion of a solute through these unstirred layers will contribute to the value of P_j calculated on the basis of the overall concentration difference, whereas the 'effective' concentration difference will be less than that used to calculate P_j. For this

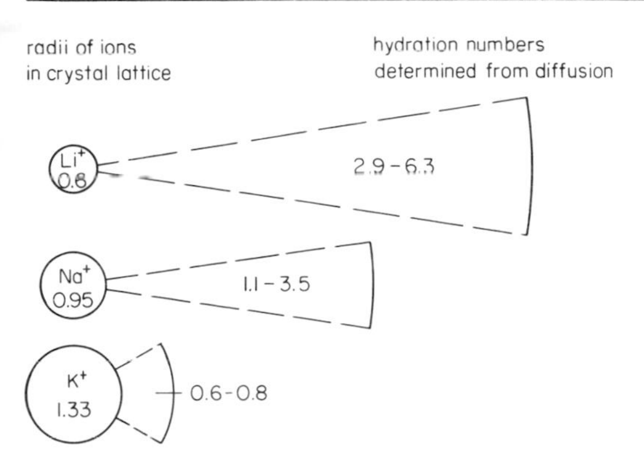

Fig. 3.2 The relative size of some alkali cations. The radii of ions can be determined accurately from measurements of crystal lattices. The actual size of the hydration shells cannot be determined unambiguously or in absolute units. The numbers and the radii shown are therefore only relative.

reason determinations of P_j require vigorous stirring of the bathing medium to reduce the thickness of the unstirred layers – but they cannot be completely eliminated. Under experimental conditions with vigorous stirring the unstirred layer may still have a thickness of 10–100 μm which is considerably thicker than biological membranes (see p. 17). In some cases diffusion through the unstirred layer can become the rate-limiting factor for some solute molecules which rapidly penetrate the membrane itself. For convenience and simplicity we will limit further treatment of concentration differences to the overall different $(c_j^i - c_j^o)$ remembering that this is an overestimate of the effective concentration difference across the membrane.

Typical values of P_j for biological membranes for small non-electrolytes such as isopropanol or phenol are about 10^{-6} m s^{-1}, while for small ions such as potassium and sodium values of about 10^{-9} m s^{-1} are usually obtained. As pointed out above, this difference in the permeability of electrolytes and non-electrolytes is mainly due to the much lower partition coefficients of electrolytes. The permeability of membranes to water P_w has also been measured and found to be about 10^{-4} m s^{-1} for many cell membranes. However, it is difficult to measure P_w with any great accuracy because of the greater relative importance of the unstirred layers in this instance.

The monovalent alkali cations form a series in which the mass and charge on the atomic nucleus for unhydrated ions increase with increasing atomic number in the following order: $Li^+ < Na^+ < K^+ < Rb^+ < Cs^+$; hydrated ions form a series which increases in the reverse order: $Cs^+ < Rb^+ < K^+ < Na^+ < Li^+$. This difference may be attributed to the fact that water molecules which are close to the charged nucleus are bound more firmly and thus the hydration radius of the ion will be greater. Thus small atoms yield large hydrated ions (see Fig. 3.2). Absolute values of the hydrated radii of ions are difficult to estimate, but relative sizes are reflected in the observed values for *ionic mobilities* obtained when an electric field is applied to a solution. For instance, the mobilities of potassium, sodium and lithium at 25° C are 7.62, 5.19 and 4.01 $m^2 s^{-1} V^{-1}$, respectively: the smaller the non-hydrated ion the lower is the mobility. In addition to the primary hydration shell, the electric field of an ion results in a secondary shell of partially structured water.

There are at least two possible ways in which an ion can cross a biological membrane: (1) by losing its associated water and moving into solution in the membrane, and (2) by passing through hydrophilic holes or pores in the membrane which are sufficiently large to accommodate an hydrated ion. There are points in favour of, and also against, these two possibilities. The activation energy for ion permeation is not high enough to include the energy which would be involved if dehydration were a prerequisite for ion movement through the membrane. The presence of pores in the membrane has been suggested to account for the high permeability to water. Other evidence for pores is provided by the values of the reflection coefficient σ which have been obtained for a number of substances and membrane systems. For example, σ for urea in intestinal epithelial cells is 0.8, while for lactose it is 0.97. The molecular radius of urea is 0.23 nm and that of lactose 0.54 nm. It may be, therefore, that there are pores with a radius smaller than 0.54 nm which allow urea but not lactose to pass. It is also true that in many cases the rate of permeation of monovalent cations correlate with the hydrated ion radius, although notable exceptions occur. For example, potassium permeates faster than sodium, but in a few cases, such as the outside of the frog skin, sodium permeates faster. This suggests that a simple sieving action by pores is not sufficient to account for the observed selectivity of membranes. It has therefore been necessary to postulate the action of carrier molecules which are more soluble in the lipid membrane phase than the ions they carry, but which allow the ions to be sequestered from the hydrophobic interior of the membrane during their passage between the aqueous phases at either surface.

Carriers

The concept of a facilitated diffusion of substances across biological membranes has been introduced elsewhere (p. 51). Facilitated diffusion systems can be identified by certain criteria:

1. The rate of transport of a substance down its gradient of electrochemical potential is more rapid than that predicted from its molecular size and lipid solubility.
2. Fick's law is not obeyed.

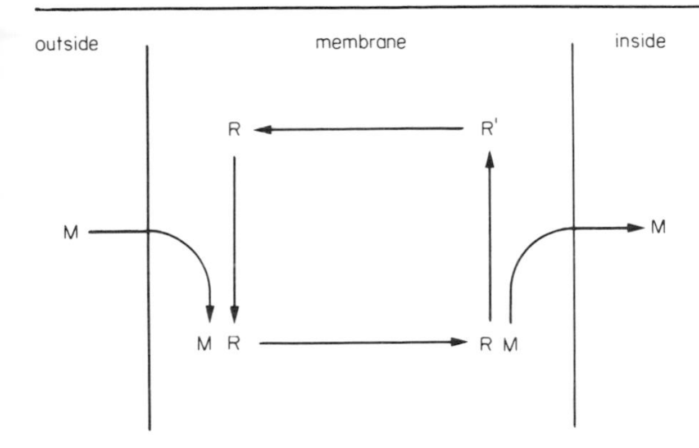

Fig. 3.3 The carrier hypothesis. M is the penetrating ion, R the carrier, and MR the mobile complex. R′ is the mobile carrier precursor.

3. Competition occurs between structural analogues.
4. Denaturing of the carrier depresses the rate of transport.
5. Counter-transport in the opposite direction may take place simultaneously.

Such facilitated diffusion requires a combination with a *carrier molecule* within the cell membrane which is able to traverse the membrane, release the substance and then return unloaded. This simple carrier hypothesis is illustrated in Fig. 3.3. A penetrating ion M combines with the carrier R at the outer surface of the membrane. This stage might involve adsorption, exchange adsorption or some kind of chemical reaction. The complex MR cannot leave the membrane but is mobile within it and may move to the inner side of the membrane, where it is broken down, releasing the ion and forming a carrier precursor R′. This precursor then moves back across the membrane and is reconverted to R which can now accept another ion at the surface. Thus a limited number of carrier molecules are capable of transporting an indefinite number of ions.

A widely held view is that the ion carrier compounds are proteins and probably enzymes. Proteins have a number of properties which are suited to such a role in ion transport. They are invariably constituents of biological membranes (see p. 11); they are capable of combining reversibly with specific ions; and they are able to assume various configurations, thus altering their shape and position in the membrane from time to time. Furthermore, proteins which are capable of binding various substances have been isolated from bacterial membranes, but it has not been conclusively established that they are the functional carrier system within an intact membrane (see p. 13). For instance, in the case of sulphate transport a pure, crystalline protein of molecular weight 32,000 which will bind sulphate has been isolated from membranes of the bacterium *Salmonella*

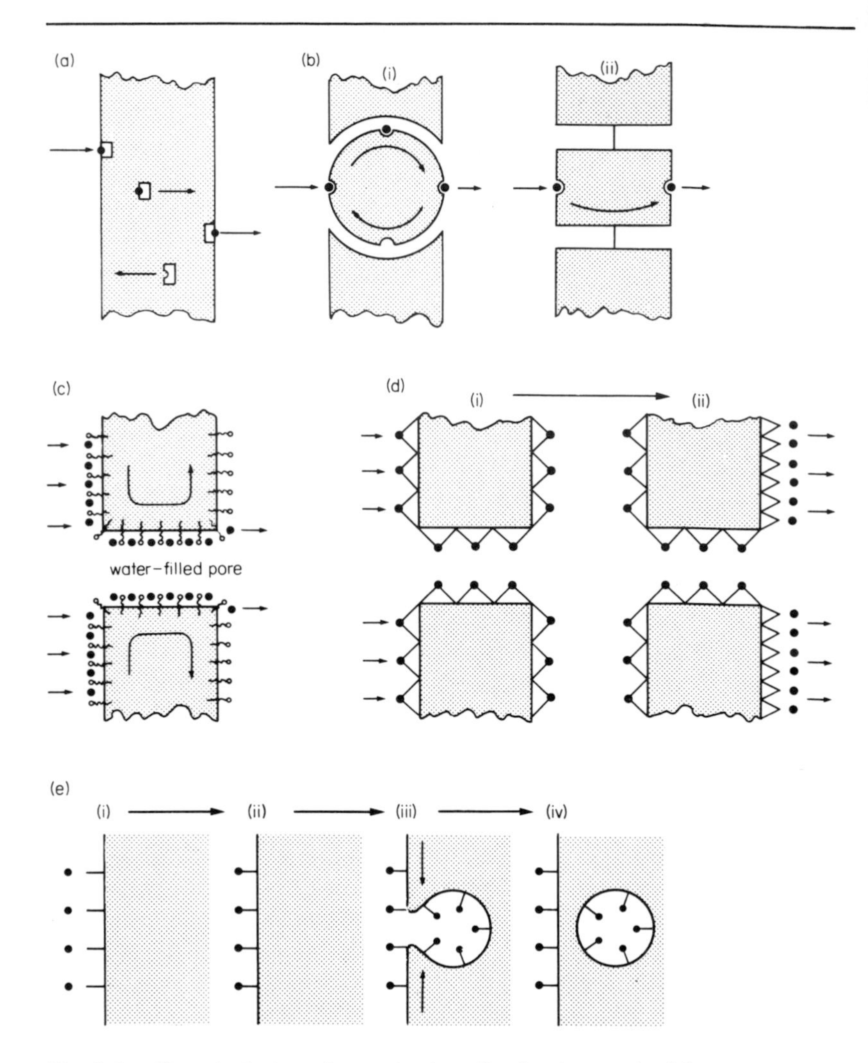

Fig. 3.4 Hypothetical carrier mechanisms for the transport of ions across membranes. (*a*) Diffusing shuttle; (*b*) rotating carriers; (*c*) sliding carrier; (*d*) contractile carrier (after Danielli, 1954); (*e*) pinocytosis (after Bennett, 1956).

typhimurium (Pardee and Prestidge, 1966). However, mutants which are unable to transport sulphate have the same protein in their membranes, and thus the protein may only be part of the carrier system.

In macromolecular terms, carrier transport, whether as facilitated diffusion accelerating the movement of molecules down a gradient, or as active transport where such movement is against a gradient, requires some change in the macromolecular configuration of the carrier within the membrane. Danielli (1954) has discussed the possible mechanisms involved in such transport, details of which are presented in Fig. 3.4. The simplest mechanism is that of the *diffusing shuttle* (Fig. 3.4*a*), the essentials of which are that the carrier should be lipid-soluble and relatively water-insoluble, and should form a complex with the transported substance readily and reversibly, with both carrier and complex diffusing readily in the lipid layer. Such carrier molecules could effectively shuttle across the membrane, accelerating the attainment of equilibrium, i.e. they would bring about facilitated diffusion. Another possible mechanism is that of a *rotating carrier* (Fig. 3.4*b*) which transports ions across the membrane by the rotation of all or part of the carrier molecule. The binding site or sites would then return to the outer surface unloaded. Alternatively, the carrier might be a substance which is strongly surface-active, able to slide along the membrane surface and through water-filled pores, the polar head with its bound ions being located in the water phase and the lipophilic tail within the membrane lipid phase (Fig. 3.4*c*).

In addition to those mechanisms which would facilitate passive diffusion, other mechanisms more directly dependent on cellular metabolism have been proposed. *Contractile proteins* may function as active carriers (pumps) transporting ions as a result of rhythmic contraction and expansion of the polypeptide chain (Fig. 3.4*d*). Micro-vesiculation of a part of the membrane to which ions have become bound may result in *pinocytotic vesicles* which break down and release the ions once across the membrane (Fig. 3.4*e*). Another possible result of changes in protein configuration is the production of transient pores in the membrane by energy-dependent processes. Such a mechanism would account for the porous behaviour of membranes under certain circumstances (see p. 62).

Ionophores

In recent years the study of ionophoric substances has provided a clue to the molecular basis of carrier transport of ions across membranes. These ionophores are antibiotic molecules produced by bacteria and fungi which have been observed to increase greatly the permeability of artificial lipid bilayers and biological membranes to ions, in some cases in a selective manner. These substances are effective even when present at very low concentrations (10^{-8}–10^{-3} mol m^{-3}), and it has been speculated that molecules with similar properties may be the carriers in biological membranes. Antibiotics such as valinomycin have been shown to increase the permeability to potassium of membranes of mitochondria, red blood cells and synthetic lipid bilayers by factors of several thousand, often with little effect on permeability to sodium.

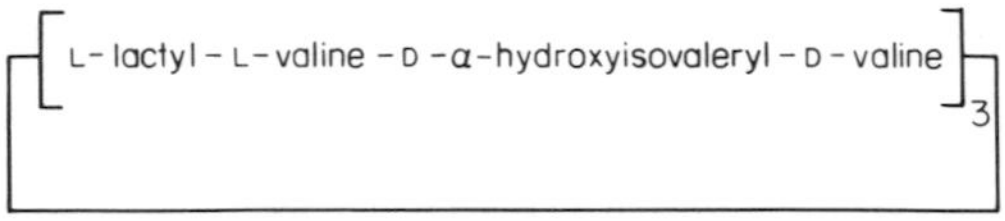

[L-lactyl - L-valine - D-α-hydroxyisovaleryl - D-valine]$_3$

valinomycin

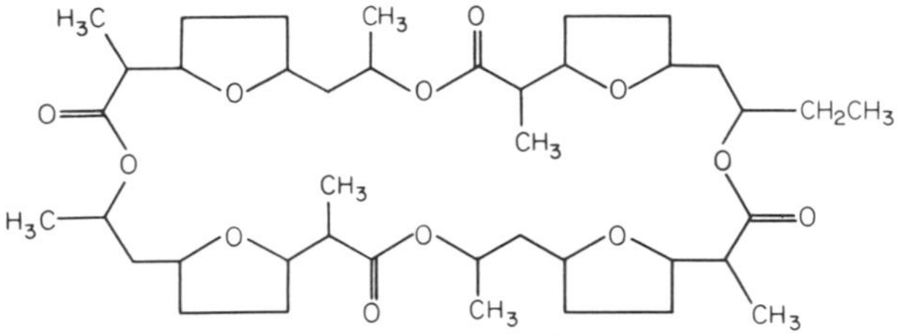

monactin

(a) the cyclic dodecadepsipeptide valinomycin and the macrotetrolide monactin

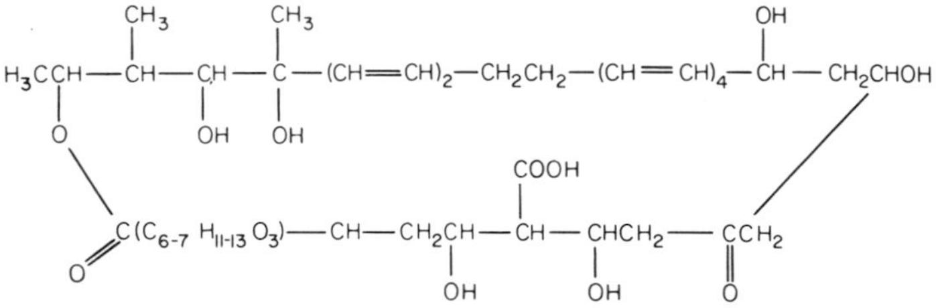

(b) the approximate structural formula for nystatin

Fig. 3.5 Structural formulae of (*a*) the carrier-type ionophores, valinomycin and monactin, and (*b*) the pore-forming ionophore, nystatin.

There appear to be two ways in which ionophores act. They either form a complex with the ions transported which diffuses through the lipid phase of the membrane in a facilitated manner, or they induce the formation of transient pores in the membrane through which ions enter. Ionophoric effects appear to be restricted largely to cations and protons although some anion selective molecules have been found (see below).

Valinomycin and *monactin* are two ionophores which form lipid-soluble complexes with cations which result in the rapid transport of the ions across membranes. These two ionophores are both macrocyclic structures (Fig. 3.5*a*). The chemical groups situated on the outside of the ring are non-polar, accounting for the observed lipid-solubility of the molecule. On the inside of the ring the oxygen atoms provide a polar environment with sites for the electrostatic binding of cations. Thus an ion in this exchange site could cross the membrane surrounded by a hydrophobic layer.

The pore-forming ionophores are different in structure, as illustrated by *nystatin* (Fig. 3.5*b*), a polyene which shows appreciable anion selectivity. This molecule is also macrocyclic, but polar groups project from both the outer and the inner faces of the ring structure. Because of this feature it is unlikely that the observed ionophoric properties of nystatin are due to the formation of ion-carrier complexes, and evidence has accumulated which suggests that nystatin opens up pores in the membrane. For instance the phospholipid bilayer is normally impermeable to urea (radius 0.23 nm), glycerol (radius 0.27 nm) and glucose (radius 0.40 nm), whereas on addition of nystatin both urea and glycerol can cross the membrane, but not glucose (Holz and Finkelstein, 1970). This observation suggests the formation of a pore smaller than 0.40 nm, which is very close to the value predicted on p. 62 in considering the magnitude of the reflection coefficient for membranes.

The hole in the nystatin ring structure also has a radius of approximately 0.4 nm, but this result is probably fortuitous as the molecule is not large enough to span the lipid bilayer and, as stacking of the molecules is unlikely, it would appear that the pore is formed in some other manner. One possibility is that the nystatin interacts with the cholesterol in the lipid bilayer, thereby inducing the formation of lipid micelles which, when closely packed together, result in hydrophilic pores with a radius of approximately 0.4 nm.

It is of great significance that an artificial, lipid bilayer membrane with ionophores of the two types outlined above would closely approximate the properties of a biological membrane, thereby providing both mobile cation-selective carriers and anion-selective water-filled pores. Finally, it should be mentioned that the principle on which ionophores selectively mobilize ions may be based on the general theory of membrane selectivity described by Eisenman (1962) and discussed in the next section.

Membrane selectivity

The observation that biological membranes discriminate between different cations raises the question as to which property of the membrane causes this selectivity. Studies by Eisenman on the cation-selective properties of glass electrodes and ion-exchange resins have indicated that for the five alkali cations Li^+, Na^+, K^+, Rb^+ and Cs^+ only 11 selectivity orders are found (Table 3.1), whereas there are 120 possible selectivity sequences. Two limiting selectivity orders are obtained: I, that in which the non-hydrated molecular radii decrease, and XI, the converse order in which the hydrated molecular radii decrease, with 9 intermediate sequences. Similar

Table 3.1 Selectivity orders found for the binding of alkali cations by non-biological materials (see text).

No.	Order
I	$Cs^+ > Rb^+ > K^+ > Na^+ > Li^+$
II	$Rb^+ > Cs^+ > K^+ > Na^+ > Li^+$
III	$Rb^+ > K^+ > Cs^+ > Na^+ > Li^+$
IV	$K^+ > Rb^+ > Cs^+ > Na^+ > Li^+$
V	$K^+ > Rb^+ > Na^+ > Cs^+ > Li^+$
VI	$K^+ > Na^+ > Rb^+ > Cs^+ > Li^+$
VII	$Na^+ > K^+ > Rb^+ > Cs^+ > Li^+$
VIII	$Na^+ > K^+ > Rb^+ > Li^+ > Cs^+$
IX	$Na^+ > K^+ > Li^+ > Rb^+ > Cs^+$
X	$Na^+ > Li^+ > K^+ > Rb^+ > Cs^+$
XI	$Li^+ > Na^+ > K^+ > Rb^+ > Cs^+$

selectivity orders have been found in biological systems (Table 3.2), which implies that the physical basis of the selectivity is the same.

To explain these results Eisenman advanced the theory that the basis of cation selectivity in membranes or in a macroscopic phase was the field strength of the fixed electronegative sites in the system. The equilibrium cation specificity of such a site depends on the free energy differences between the ion–site and ion–water interactions, the forces involved being mainly coulombic. When a cation is hydrated or adsorbed to a fixed negative charge there is a decrease in the free energy of the system, the level of which determines the relative affinities of the binding site for a cation. The

Table 3.2 Selectivity orders found for the binding of alkali cations by biological materials.

Order no.*	Material
III	*Chlorella* ion exchange
IV	Frog sartorius muscle flux
VI	Blowfly salt receptor stimulation
IX	Squid action potential
X	Frog skin permeability (P^{oi})
XI	Cation binding of DNA and RNA

*(See Table 3.1)

cation which exhibits the greatest decrease in free energy will be selected preferentially.

The field strength of an exchanger will be increased when the sites are more closely spaced or when the radius of the fixed anion is decreased. Thus field strength is correlated with the pK (the ionization constant), increasing pK values of a site giving increased field strengths, i.e. weaker acid sites have higher field strengths. The nature of these binding sites in biological materials is limited to either carboxyl or phosphoric acid groups for cations and to amino groups for anions. However, even this restricted choice allows for a wide range of binding sites, which will differ in field strength, and thus in ion-selectivity, as outlined above.

The selectivity of sites for anions and for divalent cations can also be predicted by a similar coulombic model, the relative affinities of sites for monovalent or divalent cations being dependent on site spacing, close spacing favouring divalent cations and wide spacing favouring monovalent cations.

The selectivity of ionophores can also be explained on the basis of the theory outlined above, and indeed monactin has been shown to have the following selectivity series (Eisenman *et al.*, 1968):

$$K^+ > Rb^+ \gg Cs^+ > Na^+ \gg Li^+ .$$

It is unfortunate that proton exchange is not covered by Eisenman's theory, and the difficulty of accurately determining proton exchange limits the application of the theory to biological systems.

Membrane potentials

The membrane potential E_M is the result of the diffusion and active transport of ions across the plasma membrane with a further contribution from the synthesis of negatively charged proteins. The diffusion potential across the plasma membrane arises because permeating ions of opposite charge tend to move at different rates. As electrical neutrality is maintained within the system, charge separation does not occur (i.e. there is no movement of electrical current across the membrane) since the faster moving ions are slowed down by slower moving ions of opposite charge until they both move at the same rate. Thus there is no flow of electric current, but an electric potential termed the *membrane diffusion potential* is generated.

The active transport of ions across the membrane may also contribute to the membrane potential. If such active transport is by a *neutral ion pump* it will not contribute directly to the electric potential but will result in an asymmetric distribution of the ion transported across the membrane by the pump. If for instance potassium is transported inwards and sodium outwards by a linked pump, the increasing asymmetry of potassium across the membrane will result in the term ln $(P_K c_K^o / P_K c_K^i)$ becoming more negative in the Goldman constant field equation, [2.10]. The term $\ln (P_{Na}c^o_{Na}/P_{Na}c^i_{Na})$ will become more positive, but as $P_K > P_{Na}$ for most membranes the resultant ionic asymmetry will cause an electric potential. If the neutral ion pump is inhibited, potassium will leak out and the value of E_M will gradually fall to a lower, equilibrium value.

If an ion pump is *electrogenic*, charge is moved in one direction only during the ion transport. The effect of such a transport is to increase the charge asymmetry, the resultant potential of which is additive and therefore contributes to the electrical diffusion potential across the membrane. Inhibition of such an electrogenic pump will cause abrupt changes in the value of the observed membrane potential (see p. 39).

Protein molecules within the cytoplasm carry a net negative charge at the prevailing pH and thus provide large anions which cannot move across the plasma membrane. This results in an asymmetry of mobile ions across the membrane to give an equilibrium, termed the *Donnan equilibrium*, in which the distribution of mobile anions and cations is unequal such that for potassium and chloride.

$$\frac{c^i_K}{c^o_K} = \frac{c^o_{Cl}}{c^i_{Cl}} \, . \qquad [3.4]$$

This immobility of the protein anion will cause an increase in c^i_K and a decrease in c^i_{Cl} relative to c^o_K and c^o_{Cl}, which will give rise to an electric potential of negative sign in the same manner as described above for the origin of the diffusion potential. The principles of the Donnan equilibrium and resultant electric potential, the *Donnan potential* E_D, are presented in Fig. 3.6. The relationship may be expressed as

$$E_D = \frac{RT}{F} \ln\left(\frac{c^o_K}{c^i_K}\right), \qquad [3.5]$$

an expression identical with the Nernst equation, [2.5].

It should perhaps be stressed that the membrane potential of a living cell is always ultimately dependent on metabolic energy, the passive gradient for one ion being dependent on the active transport of another, or on the synthesis of charged proteins which is also energy dependent.

Cytoplasmic selectivity

Within the cytoplasm, ions are present in solution and bound to macromolecules, thereby creating asymmetries of ionic distribution in addition to those due to selective accumulation in organelles. It is therefore difficult to know what value should be used for calculating equilibrium potentials across the plasma membrane. Some investigators have attacked the mem-

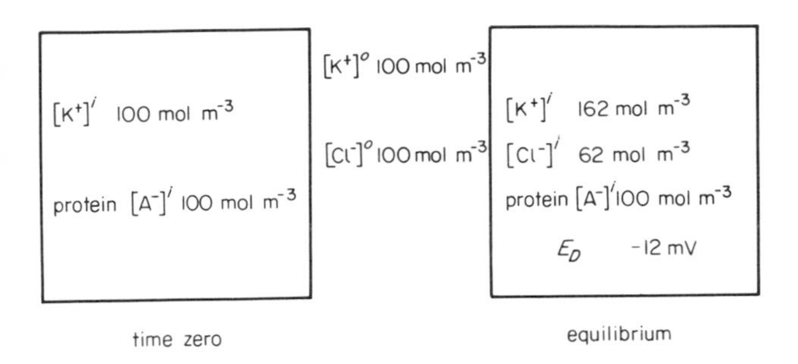

Fig. 3.6 Donnan equilibrium of ions and the Donnan potential E_D in a cell containing an indiffusible anion associated with protein. It is assumed that the plasma membrane is permeable to potassium and chloride.

brane concept as the basis for selective ion transport and have claimed that binding sites within the cytoplasm are the basis of the observed selectivity (Troshin, 1961; Ling, 1962, 1965, and 1966). These binding sites are postulated to have a marked preference for potassium over sodium, thus accounting in many instances for the observed high potassium/sodium ratio of the cytoplasm. The bound potassium would therefore have a much lower chemical activity a_K than the measured concentration c_K, which is normally used for determining the equilibrium potential (see p. 44). If the observed selectivity is the result of the above *sorption theory,* electrochemical potential gradients need not exist and the whole system would be in equilibrium. Such a concept, which is in complete contrast to the membrane theory, has not gone unchallenged by those investigators (the vast majority) who advocate that the function of the membrane is as a selective barrier.

Ion-selective microelectrodes have been used to measure the activity coefficient of ions γ_j in the cytoplasm of a number of organisms. In general γ_K is high, approaching unity, while γ_{Na} is low (<0.5), and thus the cytoplasm would appear to bind sodium preferentially rather than potassium, in direct contradiction of the sorption theory. Ling (1969) has claimed that insertion of the ion-selective electrode damages the lattice of cytoplasmic fixed charges and releases potassium in that localized region. However, if this were the case, the measured value of a_K would decrease with time, but no such reduction is observed, even after several hours. The possibility of such localized gradients of potassium can probably therefore be discounted, and the results obtained with ion-selective electrodes can be accepted as invalidating the sorption theory.

Further reading and references

Further reading

DIAMOND, J. M. and WRIGHT, E. M. (1969) Biological membranes: the physical basis of ion and non-electrolyte selectivity, *Ann. Rev. Physiol.*, **31**, 581.

DAVIES, M. (1973) *Functions of Biological Membranes.* Chapman and Hall, London.

HAYDON, D. A. (1970) The organization and permeability of artificial lipid membranes, in *Membranes and Ion Transport,* Vol. 1, p. 64, ed. E. E. Bittar. Wiley-Interscience, London and New York.

LING, G. N. (1965) The membrane theory and other views for solute permeability, distribution and transport in living cells, *Perspect. Biol. Med.*, **9**, 87.

SCHOFFENIELS, E. (1967) *Cellular Aspects of Membrane Permeability.* Pergamon, Oxford.

TROSHIN, A. S. (1961) Sorption properties of protoplasm and their role in cell permeability, in *Membrane Transport and Metabolism,* p. 45–53, eds A. Kleinzeller and A. Kotyk. Academic Press, London and New York.

Other references

BENNETT, H. S. (1956) The concepts of membrane flow and membrane vesiculation as mechanisms for active transport and ion pumping, *J. biophys. biochem. Cytol.*, **2**, 99.

DANIELLI, J. F. (1954) Morphological and molecular aspects of active transport, in 'Active Transport and Secretion', *Symp. Soc. exp. Biol.*, **8**, 502.

EISENMAN, G. (1962) Cation-selective glass electrodes and their mode of operation, *Biophys. J.*, **2**, 259.

EISENMAN, G., CIANI, S. M. and SZABO, G. (1968) Some theoretically expected and experimentally observed properties of lipid bilayer membranes containing neutral molecular carriers of ions, *Fed. Proc.*, **27**, 1289.

HOLZ, R. and FINKELSTEIN, A. (1970) Water and non-electrolyte permeability induced in thin lipid membranes by the polyene antibiotics nystatin and amphotericin B, *J. gen. Physiol.*, **56**, 125.

LING, G. N. (1962) *A Physical Theory of the Living State.* Blaisdell, New York.

LING, G. N. (1966) Cell membrane and cell permeability, in *Biological Membranes: Recent Progress.* Ann. N.Y. Acad. Sci., New York.

LING, G. N. (1969) Measurements of potassium ion activity in the cytoplasm of living cells, *Nature*, **221**, 386.

PARDEE, A. B. and PRESTIDGE, L. S. (1966) Cell-free activity of a sulphate-binding site involved in active transport, *Proc. natl. Acad. Sci. USA*, **55**, 189.

Chapter 4

Linkage to metabolism

Many of the transport processes that are essential to living cells are intimately linked to the cells' metabolism and are therefore referred to as *active transport.* The problems of defining active transport have already been discussed in Chapter 2, and we shall take here as the most practical definition that of a process by which an ion is moved against its electrochemical potential gradient (see p. 57). Such a process must therefore perform work and so consume energy; mechanisms which perform this active transport are usually referred to as *ion pumps.* Such pumps are able to maintain a constant cellular ionic composition and concentration, which may be very different from that in the external medium. To achieve this they have the ability to extract and concentrate particular nutrients which may be present in the environment at extremely low levels. For example, the cells of the body contain potassium at a higher, and sodium at a lower, concentration than that in the surrounding fluid, resulting in gradients of sodium and potassium of about 15:1 or higher. Another example is provided by the gastric mucosa in which the cells have an internal proton concentration of about 10^{-4} mol m^{-3} but secrete gastric juice with a proton concentration of 100 mol m^{-3}. Plant cells show a similar impressive ability to select and concentrate ions (Fig. 4.1 and Table 1.1); a large part of these ions is accumulated in the vacuole. Epstein (1972) has described this process as 'mining' of the environment, pointing out that only green plants and some micro-organisms have the ability to extract and accumulate ions efficiently from the environment external to the organism. However, it must be stressed that the demonstration of accumulation referred to above does not prove that active transport is occurring. A substance may be bound to cellular components on entry into the cell and so may not move against an overall electrochemical potential gradient.

Ion pumping must be a very basic biological process, utilizing a considerable portion of the cell's available energy to move ions 'uphill' in a thermodynamically unfavourable direction and so maintain these concen-

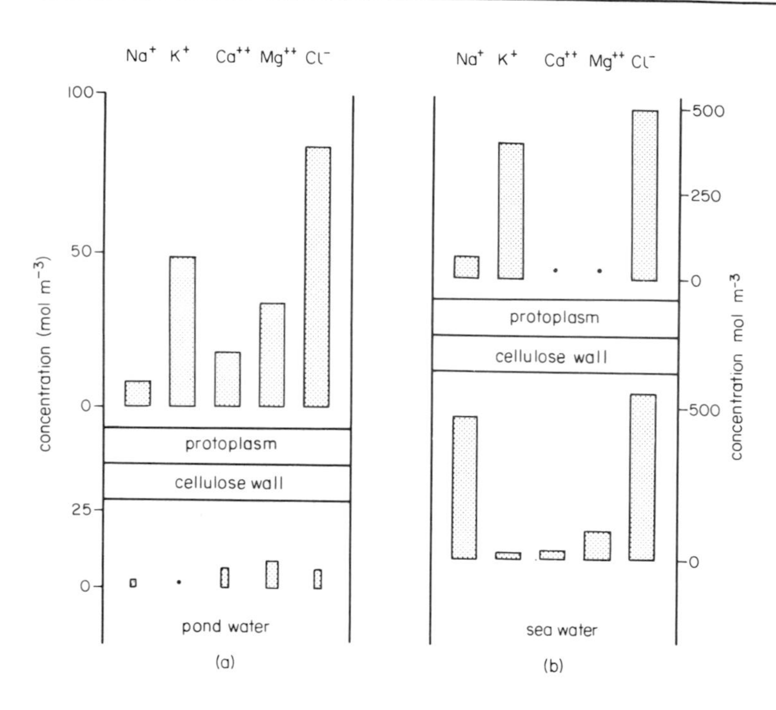

Fig. 4.1 Diagrammatic representation of the concentrations of various ions in the sap of the plants and in the medium in which they were grown: (*a*) *Nitella clavata* and (*b*) *Valonia macrophysa.* (Redrawn from Hoagland, 1944.)

tration gradients. It has been calculated that as much as a third of the energy production of an animal cell may be associated with active transport of sodium and potassium. This massive expenditure of energy is necessary since in thermodynamic terms cells are open systems and so require a continual supply of energy for their survival. In contrast, isolated systems move to a state of maximum entropy or disorder as predicted by the second law of thermodynamics. Living organisms survive by drawing a continual supply of energy or negative entropy (e.g. in the form of complex highly ordered food molecules) from their environment, a process described by Schrödinger as feeding on negative entropy.

With an overall concept of active ion transport, we must now consider the mechanisms by which energy is supplied to various active transport systems. First it is necessary to distinguish between *primary* and *secondary active transport.* In the former, the energy required for transport is supplied directly by a chemical reaction involving the transport system. As we shall

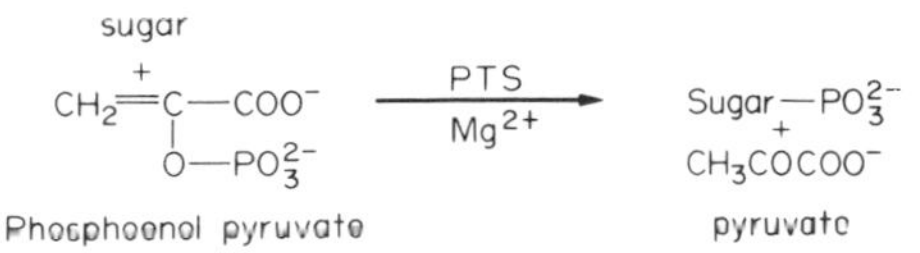

$$\text{sugar} + \underset{\text{Phosphoenol pyruvate}}{CH_2{=}C(O{-}PO_3^{2-}){-}COO^-} \xrightarrow[Mg^{2+}]{PTS} \text{Sugar}{-}PO_3^{2-} + \underset{\text{pyruvate}}{CH_3COCOO^-}$$

Fig. 4.2 The overall reaction for the phosphoenol pyruvate-dependent phosphotransferase system (PTS), an example of group translocation.

see later, this energy comes from the direct enzymatic splitting of covalent bonds, usually of ATP, or by coupling to certain redox reaction systems such as those occurring in the chloroplasts and mitochondria. In secondary transport, the uphill movement of one solute is coupled to and driven by the flow of a second solute down its gradient of electrochemical potential, this process being referred to as *co-transport* (see p. 114).

Before considering the nature of this energy coupling to ion transport in more detail, the process of *group translocation* will be briefly considered, since it is a phenomenon of related biological interest. The term transport, whether passive or active, assumes that the solute passes from one phase to another, re-appearing in the second phase in the same state. In group translocation the form appearing in the second phase differs from that leaving the first as a result of a covalent change in the transported substrate which is a part of the transport system. The carrier molecules behave as enzymes in that they catalyze group transfer reactions and, although this may result in accumulation, the solute is altered. This process may be illustrated by considering one of the best documented examples, the phosphoenol pyruvate-dependent phosphotransferase system (PTS), which mediates the uptake of sugars in a large number of bacteria. The overall reaction for this process is shown in Fig. 4.2.

The term PTS in the figure equation represents a number of protein components, one of which is tightly bound to the membrane (enzyme II) and two more of which are found in the cytoplasm (enzyme I and a protein designated HPr). The transfer of the sugar involves two basic steps: in the first step enzyme I phosphorylates HPr at the expense of phosphoenol pyruvate, while in the second step this phosphorylated protein is used as a phosphate donor for the sugar in a reaction catalyzed by enzyme II. It is believed that this system has a binding site for sugar on the outer face of the membrane and that a conformational change both phosphorylates and transfers the sugar into the cell in a single operational step (Fig. 4.3). Enzyme II probably carries the specificity for sugar recognition and there may be a different enzyme II for each sugar transported by the bacteria. This concept is supported by evidence obtained using a number of mutants which lack various species of enzyme II and as a result are unable to transport the corresponding sugars.

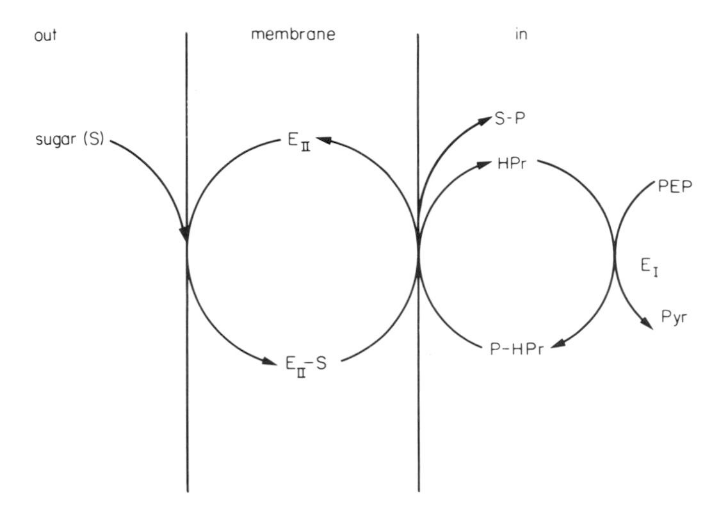

Fig. 4.3 Group translocation of sugars by the phosphoenol pyruvate-dependent phosphotransferase system. PEP, phosphoenol pyruvate; Pyr, pyruvate; E_{II}, an enzyme; HPr, a protein; S-P, phosphorylated sugar.

Sources of energy for transport

The two most important energy sources for primary active transport are ATP hydrolysis and electron flow connected with certain redox reactions. As both of these processes are closely associated with the membranes of the mitochondria and chloroplasts in the processes of respiration and photosynthesis, a brief description of these organelles and the reactions involved in the generation of these driving forces for transport is now presented.

Mitochondria are found in all aerobic eukaryotic cells. They are, on average, about 3–5 μm long and 0.5–1.0 μm in diameter, and consist of an inner and outer membrane, the inner membrane showing invaginations known as cristae (Fig. 4.4*a*). The bounding membranes enclose a space termed the perimitochondrial space and the inner compartment is known as the matrix. The outer membrane is freely permeable to small ions and many compounds with a molecular weight of less than 10,000, and it is therefore usually ignored as a solute barrier. The inner membrane is much more complex and has a higher proportion of protein than the outer membrane. It is osmotically responsive and is relatively impermeable to a large number of cations and anions, sugars and amino acids. As we shall see later (p. 99), the inner membrane is thought to contain specific exchange carriers known as *porters*, and there is a close relationship between the

(a)

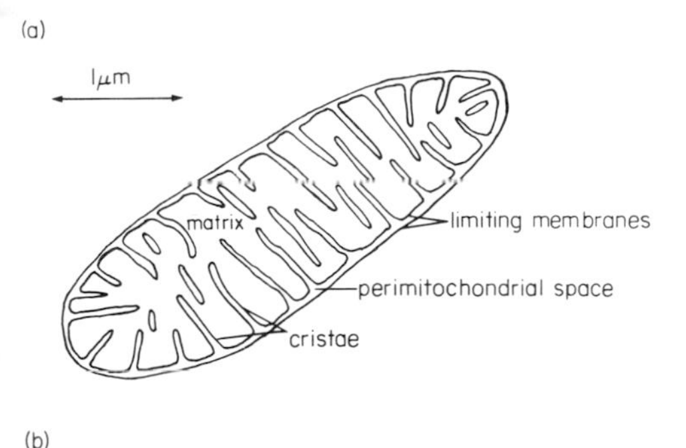

(b)

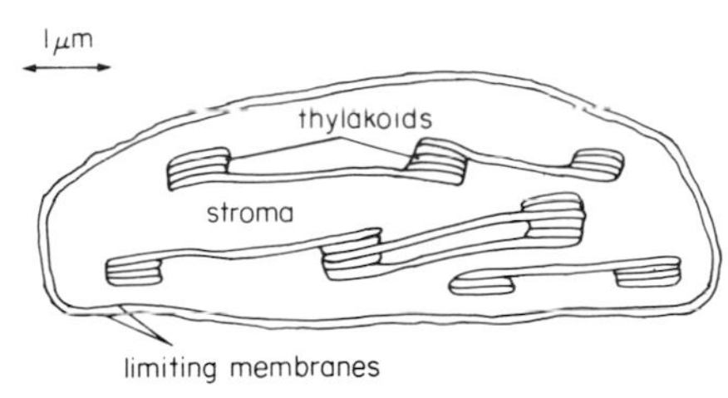

Fig. 4.4 Diagrammatic representation of the structure of (*a*) a mitochondrion and (*b*) a chloroplast.

transport properties of this inner membrane and its energy-transducing activities.

Mitochondria are referred to as the 'power houses' of the cell since they are the major sites of ATP production in non-photosynthetic, eukaryote cells. They are the location of the tricarboxylic acid cycle (or Krebs' cycle) in which pyruvate, largely produced by glycolysis, is oxidized to give carbon dioxide and hydrogen atoms in the form of reduced coenzymes. These hydrogen atoms (or their equivalent in electrons) are fed into the *respiratory chain*, which consists of a series of electron carriers of decreasing redox potential terminating in the reduction of molecular oxygen to water. This decrease in redox potential means a decrease in free energy. At three specific sites along the chain this energy is conserved by the conversion of ADP to ATP in the process of *oxidative phosphorylation* (Fig. 4.5). As we shall discuss shortly (p. 80), this process is similar in many respects to that of photosynthetic phosphorylation in the chloroplasts. A more detailed account of both of these processes is given in Hall, Flowers and Roberts (1974).

The nature of the coupling reaction between electron transport and ATP generation has been the subject of considerable controversy over a number of years. Currently, there are two major theories. One, known as the *chemical coupling hypothesis*, considers that energy transfer occurs

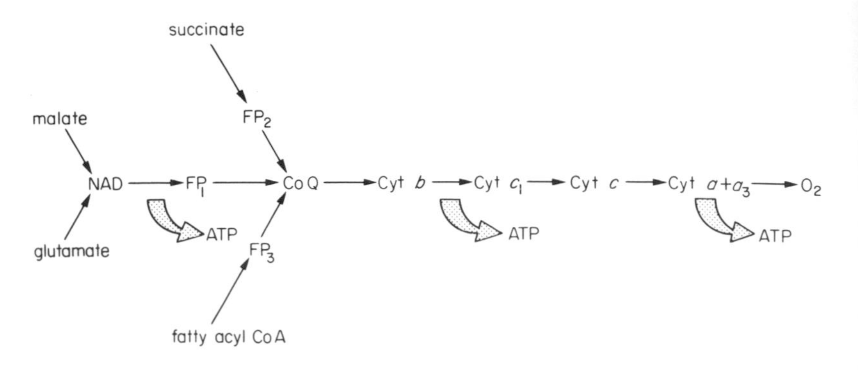

Fig. 4.5 The respiratory chain showing the entry points of electrons from various substrates and the probable sites of energy conservation by ATP formation. FP, flavoprotein; CoQ, coenzyme Q; Cyt, cytochrome. Cytochromes *a* and a_3 are together referred to as cytochrome oxidase.

through a series of common chemical intermediates which possess high-energy bonds, the final one being the precursor of the high-energy phosphate bond of ATP (Fig. 4.6). The chief alternative to this concept is the *chemiosmotic hypothesis* proposed by Mitchell (see Mitchell, 1966, 1970; Greville, 1969), which postulates that there are no common chemical intermediates between the respiratory chain and ATP synthesis. Since this latter theory has important implications to the problem of ion transport, we will consider it in greater detail.

Mitchell proposes that the electron carriers in the inner mitochondrial or chloroplast membranes are arranged *vectorially* and that the flow of electrons down this chain generates a gradient of protons across this membrane. If an enzyme has a vectorial arrangement in the membrane it is able to accept a substrate from one side and release the product at the other. In relation to the mitochondrial membrane, it is postulated that two hydrogen atoms are removed from a reductant RH_2 at one side and, after a charge separation within the membrane, protons and electrons are ejected on different sides. This generates a proton gradient across the membrane, while the electrons are able to reduce an oxidant S to SH_2. In the final stage of electron transport the oxidant S is ½ O_2. This general scheme for proton pumping is outlined in Fig. 4.7.

According to the chemiosmotic hypothesis, the inner mitochondrial membrane is essentially impermeable to protons, and as each pair of electrons moves down the chain from NADH to oxygen, three pairs of protons are extracted from the inner matrix of the mitochondria and ejected into the outer medium (Fig. 4.8). The proton gradient generated by this electron flow is then used to support the synthesis of ATP from

$$A_{red} + B_{ox} + I \rightleftharpoons A_{ox} \sim I + B_{red}$$

$$A_{ox} \sim I + X \rightleftharpoons A_{ox} + X \sim I$$

$$X \sim I + P_i \rightleftharpoons X \sim P + I$$

$$X \sim P + ADP \rightleftharpoons X + ATP$$

Fig. 4.6 The chemical coupling hypothesis. A and B are two electron carriers which are alternatively oxidized and reduced. Oxidation of A by B generates a high-energy linkage with a coupling factor I, which then exchanges the electron carrier for another component X. This is converted to a phosphorylated intermediate which is the precursor of the high-energy phosphate bond of ATP.

ADP and inorganic phosphate P_i. This is a condensation reaction involving the removal of the elements of water from ADP and P_i and is again considered to be a vectorial system in which H_2O is removed as H^+ and OH^- on opposite sides of the membrane, resulting in charge neutralization (Fig. 4.8).

There is now considerable evidence from studies of both mitochondria and chloroplasts to support the chemiosmotic theory, although it is beyond the scope of this book to discuss it in detail. It is, however, important to appreciate that the energy derived from the proton gradient (the *proton motive force*) may be used to drive oxidative phosphorylation *or* to provide the osmotic work needed for the accumulation of ions.

A similar system of electron transport and phosphorylation occurs in photosynthesis which, in eukaryotic plant cells, is associated with the chloroplasts. As with the mitochondria, these organelles are surrounded by two membranes which enclose a complex system of double membranes known as lamellae or thylakoids (Fig. 4.4*b*); this thylakoid system does not appear to be connected with the outer membranes. Photosynthetic bacteria do not contain chloroplasts although they frequently contain lamellae grouped to form distinct chromatophores. In both cases the photosynthetic pigments are bound to protein and/or lipid in these lamellar membranes. Chloroplasts are considerably larger than mitochondria, being about 5–10 μm long and 2–3 μm thick, and they have proved more difficult to isolate and maintain in an intact state. Two types of chloroplast preparation are normally distinguished, intact ones (known as class I) and broken ones lacking the outer membrane (known as class II). Studies of volume changes in various solutions in the dark have shown that class II chloroplasts are relatively impermeable to glucose, sucrose, mannitol and certain inorganic anions. Similar results are obtained with class I chloro-

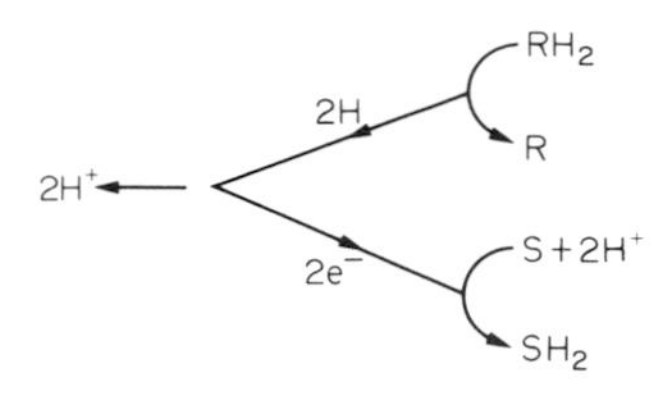

Fig. 4.7 A general scheme for proton pumping.

plasts, indicating that the outer membrane has no major role in such volume changes.

The process of photosynthesis can broadly be divided into two sequences, the *light reactions*, which are directly dependent on light, and the *dark reactions*, which may proceed independently of light but only for a short time in its absence. The light reactions involve the absorption of light and an associated electron transport system, to which is coupled the production of ATP. These reactions are of great interest in relation to considerations of the source of energy for ion transport. The dark reactions utilize the energy and reducing power generated by the light reactions to catalyze the reduction of carbon dioxide to carbohydrate. It has been demonstrated that these dark reactions are not directly associated with energy conservation and ion transport (see p. 109), and so they will not be discussed in any further detail here.

The primary reaction in photosynthesis is the absorption of light by various pigments, particularly the chlorophylls, which are arranged within the thylakoid membranes. When a light quantum is absorbed by a pigment molecule, it gains the energy of the quantum and an electron may be raised to a higher energy level, producing an excited state. Very simply, the excited electron may be passed to $NADP^+$ producing a reductant NADPH for carbon dioxide fixation and leaving a 'hole' or positive charge in the pigment molecule, which is filled by an electron from a water molecule. This sequence is known as *non-cyclic electron flow* (Fig. 4.9). Alternatively, the excited electron may be returned to its original state in the pigment molecule by moving along a series of electron carriers; this alternative pathway is known as *cyclic electron flow*. The energy of the excited electron may be conserved by the coupling of this electron flow to the phosphorylation of ADP to ATP in a process known as *photosynthetic phosphorylation* or photophosphorylation (Fig. 4.9).

In oxygen-evolving plant cells the flow of electrons in photosynthesis is more complicated, involving two light reactions, known as photosystems I and II (PS I and PS II). The most widely accepted arrangement of these photosystems and associated electron carriers is shown in Fig. 4.10. When light is absorbed by PS I, the excitation energy is transferred among the

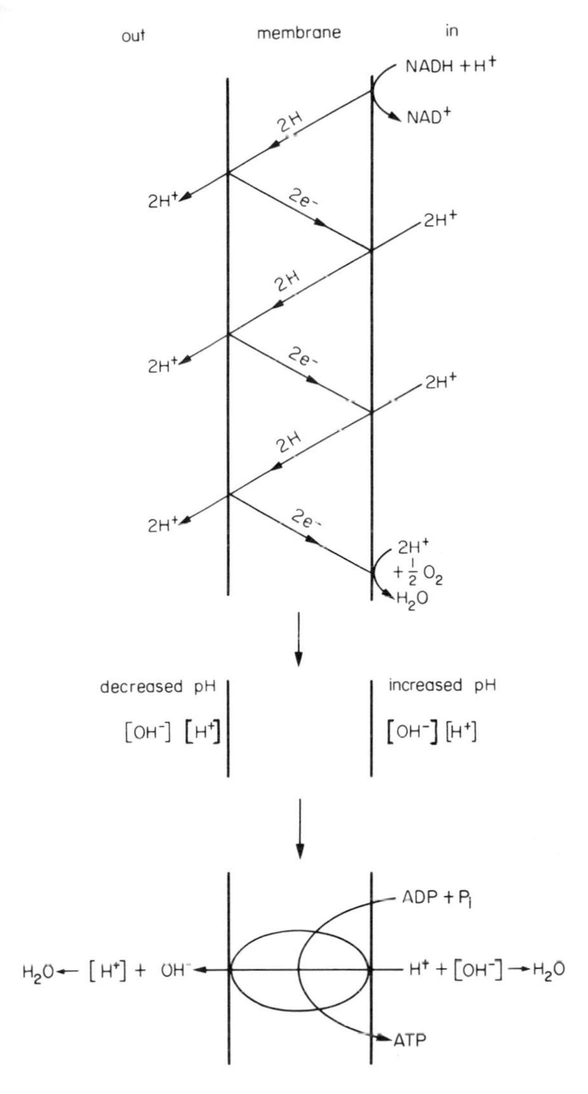

Fig. 4.8 The chemiosmotic hypothesis of oxidative phosphorylation.

chlorophyll molecules until it reaches a special chlorophyll molecule, P 700, which is referred to as a 'trap'. P denotes pigment and 700 indicates its wavelength absorption maximum. When P 700 is excited it is able to reduce ferredoxin, possibly through an intermediate, which is then in turn

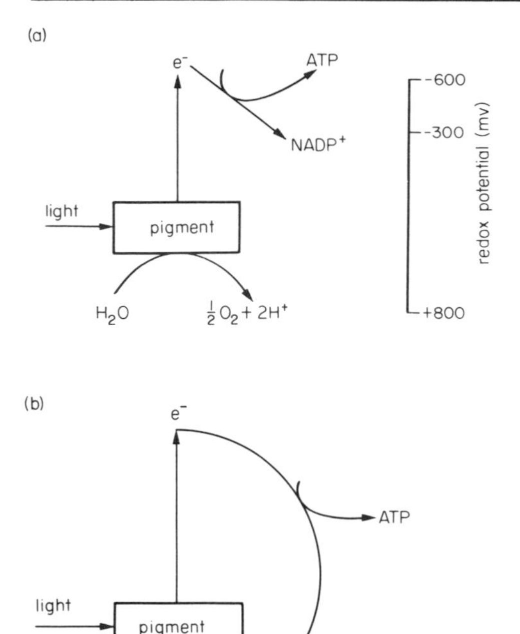

Fig. 4.9 (*a*) Non-cyclic and (*b*) cyclic electron flow. Electrons are raised to a higher energy level by the absorption of light. The electron may be transferred to $NADP^+$ through an intermediate, (*a*), or return to the electron hole left in the pigment, (*b*). This electron flow may be coupled to the phosphorylation of ADP to ATP. (Redrawn from Hall, Flowers and Roberts, 1974.)

able to reduce $NADP^+$ to NADPH. The electron hole left in the oxidized P 700 must then be refilled, and this is thought to be achieved by electrons which come ultimately from water by a process involving PS II. When PS II is excited by light, an electron is transferred to an acceptor molecule Q; the electron removed from PS II being replaced by one from water. From Q, the electron flows in an energetically downhill direction along an electron transport chain (which includes plastoquinone, two distinct cytochromes and plastocyanin) and ultimately restores P 700 to its reduced state. This flow of electrons and the associated energy drop is coupled to the phosphorylation of ADP to ATP in a process which is similar to oxidative phosphorylation in mitochondria and is termed *non-cyclic photophosphorylation.* There are probably two separate coupling sites along this electron transport chain, although their precise location is

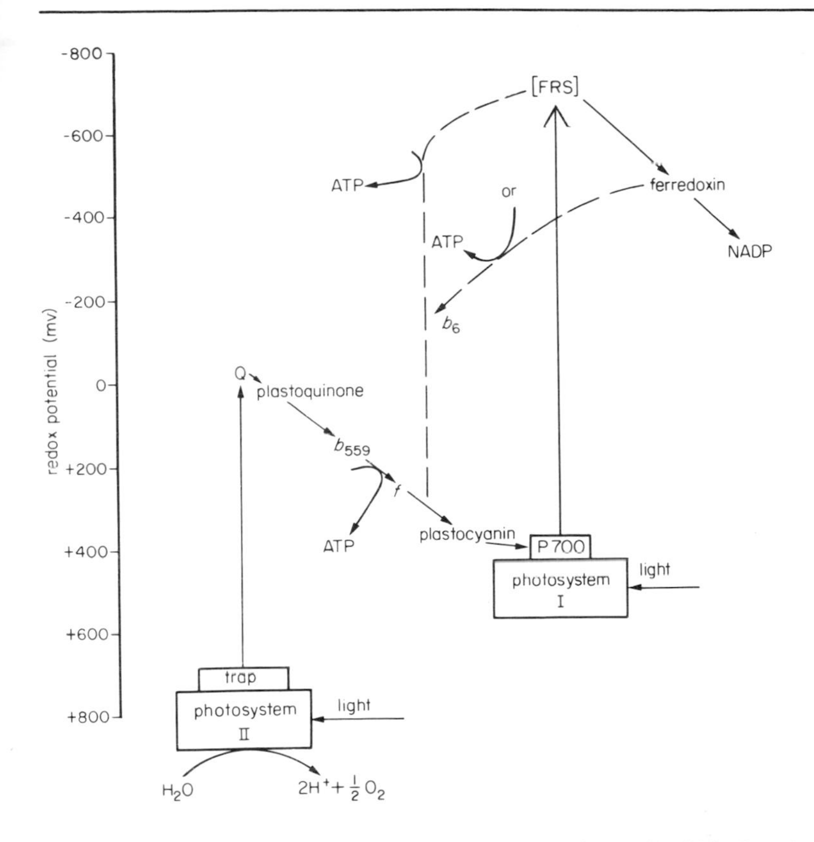

Fig. 4.10 Electron transfer between photosystems I and II. FRS, ferredoxin-reducing substance which may serve as a reductant to ferredoxin. (Redrawn from Hall, Flowers and Roberts, 1974.)

uncertain. *Cyclic photophosphorylation* may also occur under certain conditions, involving only PS I; little NADPH is formed in this case. The precise sequence of electron carriers involved in cyclic photophosphorylation is not yet clear.

As with oxidative phosphorylation, there are two major hypotheses to explain the mechanism of coupling between electron transport and ATP formation in photophosphorylation, namely chemical coupling and chemiosmosis. In the former, specific proteins known as coupling factors, are thought to act as intermediates between electron flow and phosphorylation. In the chemiosmotic theory, ATP formation is postulated to result from an electrochemical gradient of protons across the thylakoid membranes, ATP being produced as a result of a proton-driven, reversible ATP-ase. A major difference between chloroplasts and mitochondria, however, is that

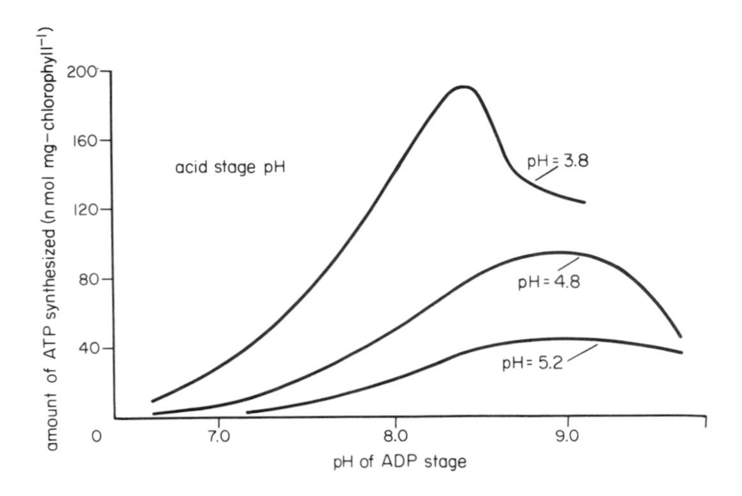

Fig. 4.11 Influence of pH changes on ATP synthesis by isolated chloroplasts in the dark. Chloroplasts were equilibrated in acid media (pH 3.8–5.2) and then transferred to media at a higher pH (6.5–10) containing ADP and P_i. This transition resulted in a synthesis of ATP, the amount formed depending on the pH gradient. (Redrawn from Clarkson, 1974, after Jagendorf and Uribe, 1967.)

protons are pumped to the outside in mitochondria but to the inside of the thylakoids in chloroplasts. The strongest experimental evidence in support of the chemiosmotic hypothesis has probably come from studies with chloroplasts. For example, if a chloroplast suspension in an unbuffered medium in the dark is illuminated, the medium becomes more alkaline before reaching a steady pH value, indicating that protons are accumulated inside the chloroplast membranes. More dramatic support is provided by the observation that pH changes in the medium can result in ATP synthesis. Jagendorf and Uribe (1967) equilibrated chloroplast suspensions at low pH values (3.8–5.2) and then rapidly adjusted the suspension to a higher pH with the accompanying addition of ADP, P_i and magnesium ions. They observed an appreciable synthesis of ATP in the dark (Fig. 4.11), a result which provides strong circumstantial evidence for the chemiosmotic hypothesis.

We have discussed above the two major sources of energy for active transport, namely ATP and a more direct coupling to electron transport. There is strong evidence that both sources are essential for ion transport in different systems, and we will now consider some of the best characterized examples of these.

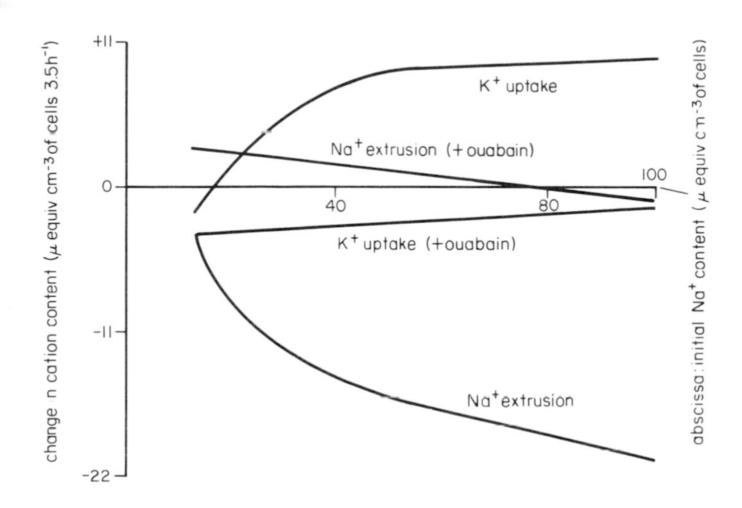

Fig. 4.12 Dependence of net sodium and potassium movements on the internal sodium concentration in human red cells. The rate of sodium and potassium movement increases with increasing internal concentration of sodium. The active movement of both ions is inhibited by ouabain which causes passive leaks in the opposite direction. (Redrawn from Whittam and Ager, 1965.)

ATP-driven transport

The sodium pump

It was stated in Chapter 1 that the ionic content of living cells frequently differs considerably from that in the external environment. This is usually the case with sodium and potassium in animal cells; the intracellular level of potassium is high and that of sodium low, whereas the reverse is found in the external medium. This discrimination is a function of the plasma membrane, which pumps sodium and potassium ions in an opposite direction to the passive leakage of sodium inwards and potassium outwards down their electrochemical potential gradients. The rate of this pumping is determined by the internal concentration of sodium, the greater the concentration the greater the rate of pumping, although this usually has an upper limit (Fig. 4.12). Thus the pump regulates its rate to meet the needs of the cell. The movement of sodium and potassium is tightly coupled, active sodium efflux requiring the presence of potassium in the external medium to be transported inwards.

This pumping is fundamental to many physiological processes. For example, many intracellular enzymes are stimulated by potassium and in-

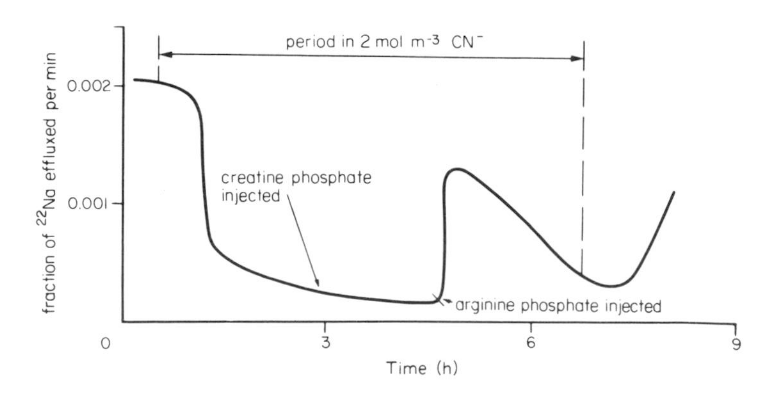

Fig. 4.13 The effect of injecting high-energy phosphate compounds on sodium efflux from a cyanide-poisoned squid axon. (Redrawn from Caldwell, Hodgkin, Keynes and Shaw, 1960.)

hibited by sodium. It is basic to various excitation processes, such as nerve conduction and muscle contraction, excitable cells responding to stimulation by allowing a rapid passive influx of sodium and passive efflux of potassium. This is the result of membrane depolarization, after which the resting membrane potential is restored by the active extrusion of sodium and the uptake of potassium. Sodium/potassium pumping is also important in various secretory processes for maintaining electrolyte and osmotic balance in a variety of epithelial tissues, such as frog skin, toad bladder, the kidney and the intestine. The active transport of sugar and some amino acids in many animal cells is probably dependent on the maintenance of sodium (and perhaps potassium) gradients across the cell membrane (see p. 114). It is now clear that the energy for the sodium/potassium pump is provided by ATP, synthesized in glycolysis or oxidative phosphorylation.

Evidence for the role of ATP

The importance of ATP in active sodium/potassium transport has been demonstrated in a number of classical experiments with squid axons and erythrocytes. Caldwell and co-workers (1960) introduced radioactive ^{22}Na into squid axons which were then poisoned with cyanide to inhibit the efflux of sodium (Fig. 4.13). The efflux of sodium could be restored by injection of ATP or arginine phosphate into the axon. ATP can be generated from arginine phosphate but not from creatine phosphate; injection of the latter had no effect on sodium efflux. Addition of these compounds to the external medium had no effect. Whittam (1958) used red cells to demonstrate the dependence of potassium influx on ATP supply. Uptake of radioactive ^{42}K into red cells was studied in the presence and absence of

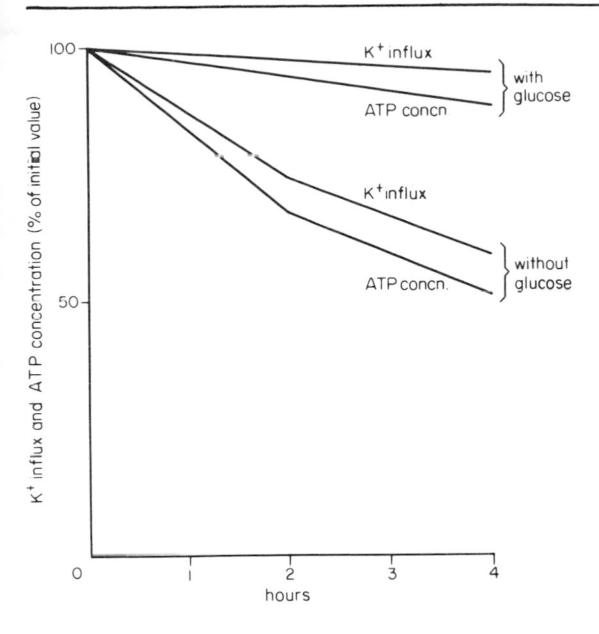

Fig. 4.14 Effect of added glucose on potassium influx and ATP concentration in human red cells. (Redrawn from Whittam, 1958.)

glucose. Without glucose, there was a steady decline in both intracellular ATP content and potassium influx. In the presence of glucose, both intracellular ATP levels and potassium influx were maintained (Fig. 4.14). Thus, these experiments and many others, have strongly implicated ATP hydrolysis as the energy source for sodium/potassium transport. Evidence for a direct coupling of ATP hydrolysis to cation pumping was provided by the discovery of a membrane-bound adenosine triphosphatase (ATP-ase) activity by Skou (1957), which required both sodium and potassium ions for maximal activity. Subsequent work on a variety of cells and tissues has confirmed the presence of a sodium–potassium activated ATP-ase (Na–K ATP-ase), and other important correlations have been observed between the enzyme activity and active sodium/potassium transport. It has therefore been proposed that this enzyme is directly involved in the transport process; it is sometimes referred to as the transport ATP-ase.

Characteristics of Na–K ATP-ase

The overall reaction catalyzed by the ATP-ase involves the hydrolysis of the γ (terminal) phosphate of ATP to give ADP and orthophosphate

$$ATP \longrightarrow ADP + P_i$$

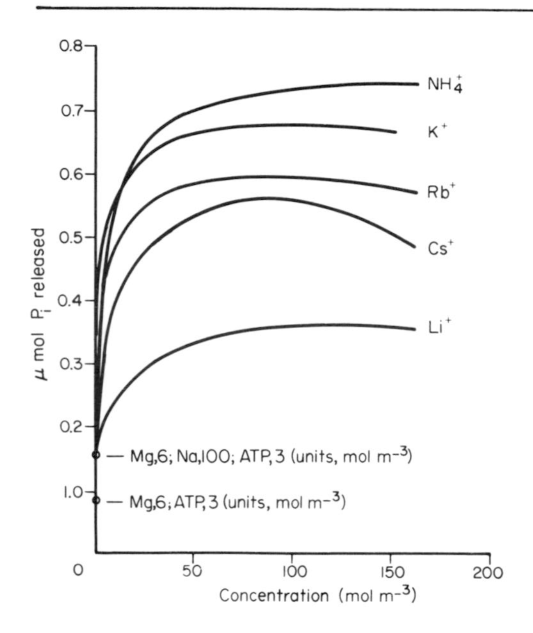

Fig. 4.15 Stimulation of crab nerve ATP-ase activity by various monovalent cations in the presence of magnesium and sodium. (Redrawn from Skou, 1960.)

The reaction is exergonic with a ΔG_0^1 value of approximately -7 Kcal mol^{-1}.

Magnesium ions are always required for this activity, although with magnesium alone the rate of hydrolysis is slow. If sodium is added, there is a small increase in activity, whereas if potassium (or another monovalent cation such as ammonium, rubidium or caesium) is added in the presence of sodium and magnesium, there is a major stimulation of activity (Fig. 4.15). Thus full activity depends on the synergistic effect of sodium and potassium. Na–K ATP-ases from a variety of animal tissues show broadly similar properties. The sodium concentration for maximum activity is usually 60–130 mol m^{-3} with an optimum sodium/potassium ratio varying between 5:1 and 10:1. Sodium is an absolute requirement but other monovalent cations can substitute for potassium.

The strongest evidence for a linkage between the ATP-ase and sodium/potassium transport has come from work on red blood cells. If the enzyme and transport systems are compared it is found that:

1. Both are integral parts of the membrane.
2. Both systems utilize ATP but not inosine triphosphate.
3. Both require sodium and potassium together and accept ammonium in place of potassium; the concentrations required for half-maximal activation are the same in both systems.

4. Both are inhibited by the cardiac glycoside ouabain (strophanthin G) at similar concentrations. Only the Na^+/K^+-stimulated part of the ATP-ase is inhibited.
5. Both systems show an effect of potassium in countering the effect of low, sub-optimal concentrations of ouabain.
6. When ATP is introduced into red cell ghosts, a close parallelism is found between potassium influx and ouabain-sensitive ATP hydrolysis; about 3 sodium ions are transported out and 2 potassium ions in for every ATP molecule hydrolyzed.

Table 4.1 Comparison of cation fluxes and Na-K ATP-ase activities. Reproduced from Bonting (1970).

Tissue	Temperature (°C)	Cation flux (10^{-10} mol m^{-2} s^{-1})	Na-K ATP-ase activity (10^{-10} mol m^{-2} s^{-1})	Ratio
Human erythrocytes	37	3.87	1.38	2.80
Frog toe muscle	17	985	530	1.86
Squid giant axon	19	1,200	400	3.00
Frog skin	20	19,700	6,640	2.97
Toad bladder	27	43,700	17,600	2.48
Electric eel, non-innervated membrane, Sachs organ	23	86,100	38,800	2.22
				2.56*

* mean value.

Finally, it may be added that these transport systems show a widespread distribution in different tissues and species and a significant correlation has been found between Na–K ATP-ase activity in a tissue and the rate of cation transport (Table 4.1). No correlation was observed between Mg ATP-ase activity and the flux values. Overall aspects of this transport process are illustrated in Fig. 4.16.

Mechanism of action

Having established the involvement of ATP-ase activity in sodium/potassium transport, it is pertinent to consider how the phosphoryl-transferring activity of the enzyme is actually coupled to ion movement. It is now believed that transport involves a phosphorylated enzyme intermediate, and it has been suggested that sodium efflux is associated with the phosphorylation step and potassium in flux with the subsequent hydrolysis. The sequence of reactions shown in Fig. 4.17 has been proposed (see Dahl and Hokin, 1974). The first step involves a reversible phosphorylation of the enzyme which is dependent on sodium and magnesium and is inhibited by high concentrations of ouabain. The second stage involves a conforma-

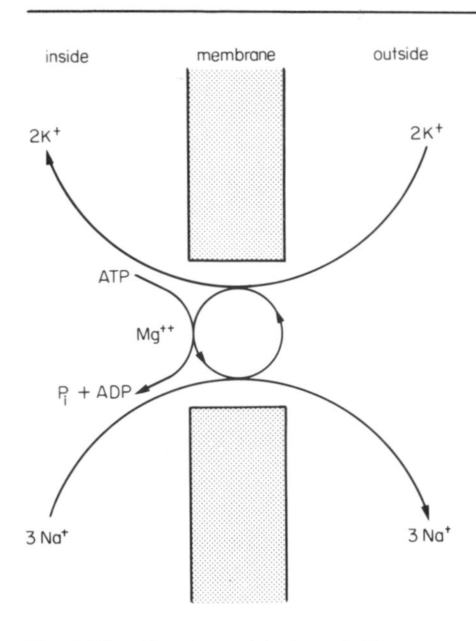

Fig. 4.16 Simple model of the ATP-dependent Na^+/K^+ transport system.

tional change in the phosphorylated enzyme intermediate which probably makes it sensitive to discharge by potassium. There is still considerable controversy concerning the exact number and nature of these phosphorylated intermediates. Finally, the intermediate is dephosphorylated by a process which is dependent on potassium and magnesium and is sensitive to low concentrations of ouabain. Ouabain can bind either with the free enzyme or the phosphorylated state to produce an inactive complex and both the rate and amount of this binding are inhibited by potassium in many mammalian cells. It should be noted, however, that this stepwise explanation of ATP hydrolysis is subject to debate. Whittam and Chipperfield (1975) have proposed that sodium and potassium are transported simultaneously with the hydrolysis of ATP in a one-step reaction.

By means of various detergents, and using tissues rich in Na–K ATP-ase activity, attempts have been made to purify the system. It appears to consist of two major proteins; one of molecular weight about 100,000 and the other about 55,000. These proteins are presumably embedded in the membrane in close contact with the lipids. It has been established that phospholipids, such as phosphatidyl choline, are necessary for activity, although their actual role in the functioning of the ATP-ase has not yet been resolved. The pump, however, is undoubtedly a complex unit of protein and phospholipid.

$$E_1 + ATP \underset{}{\overset{Na^+\ Mg^{++}}{\rightleftharpoons}} E_1{-}P + ADP$$

$$E_1{-}P \overset{Mg^{++}}{\rightleftharpoons} E_2{-}P$$

$$E_2{-}P + H_2O \overset{K^+}{\rightleftharpoons} E_2 + P$$

$$E_2 \rightleftharpoons E_1$$

Fig. 4.17 A proposed sequence of reactions by which the phosphoryl-transferring activity of the enzyme E is coupled to ion movement in Na^+/K^+ transport. (After Dahl and Hokin, 1974.)

Thus, although we know many of the steps leading to the extrusion of sodium and the uptake of potassium, we still do not know how these effects are produced. It is generally considered that cation movement involves a series of conformational changes in the Na–K ATP-ase, but it is not clear if the ATP-ase itself acts as a sodium carrier or whether the ATP-ase generates an energy-rich state which is then used to drive transport. This question will be discussed later in this chapter in relation to the general problem of energy transfer in transport (p. 101).

Regulation of metabolism by the sodium pump

We have seen that active transport of sodium and potassium in many animal cells depends on ATP. A major part of cellular metabolism may be involved in supplying energy for the pump largely from glycolysis and oxidative phosphorylation. However, this means that the pump may play an important role in the regulation of metabolism and, since the sodium pump requires external potassium to function, the level of external potassium can have an important influence on the rate of intracellular metabolism.

The importance of potassium can readily be demonstrated with mature red blood cells which, as they contain no mitochondria, rely on glycolysis for the supply of ATP. These cells convert one molecule of glucose to two molecules of lactate, and so glucose consumption can be measured by lactate production. It has been demonstrated that both lactate production and ouabain-sensitive inorganic phosphate production are influenced by the external potassium concentration (Fig. 4.18); the concentration required for half-maximal activation is close to that needed to stimulate the sodium pump. This observation strongly suggests that the influence of potassium on glucose utilization is through the sodium pump.

In more active cells, it appears that the sodium pump regulates oxygen consumption by mitochondria through the production of ADP and inorganic phosphate. For example, oxygen consumption of tissue slices of brain and kidney is related to the sodium pumping rate and is sensitive to ouabain. This relationship can be demonstrated more directly using mem-

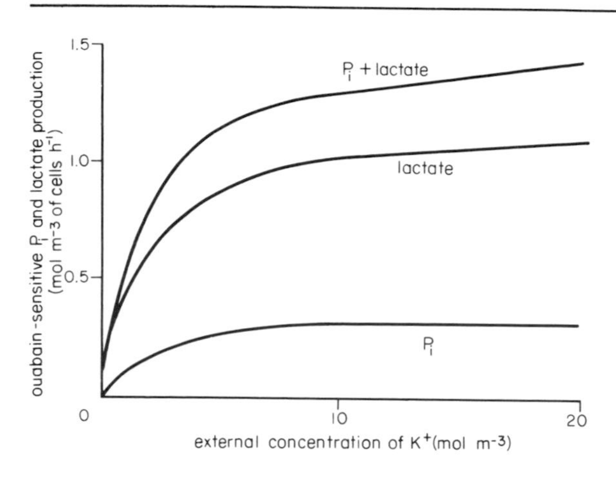

Fig. 4.18 Potassium-dependence of ouabain-sensitive inorganic phosphate and lactate production in red cells. (Redrawn from Whittam and Ager, 1965.)

brane and mitochondrial preparations from, for example, brain cortex (Whittam and Blond, 1964). Oxygen consumption by mitochondria has been measured in the presence of membranes which hydrolyze ATP at different rates in the presence of different concentrations of ouabain. A linear relationship was observed between rates of oxygen consumption and ATP-ase activity (Fig. 4.19). If internal sodium or external potassium

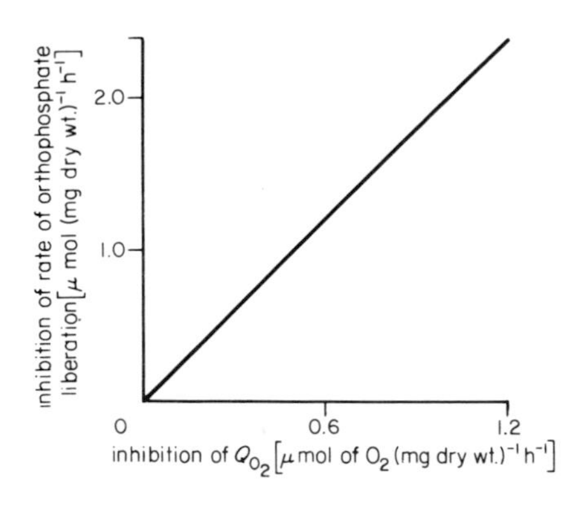

Fig. 4.19 Relationship between the graded inhibition of respiration and of ATP-ase activity in brain cortex homogenates by different concentrations of ouabain. (Redrawn from Whittam and Blond, 1964.)

was removed from the intact tissue, oxygen consumption fell by 30–40%; this presumably represents the proportion of the cell's metabolic energy supply which is utilized by the pump. Thus it seems that a considerable portion of the metabolism of many animal cells is involved in driving the sodium/potassium pump, which in turn means that the pump provides a pacemaker for metabolism.

ATP-dependent transport in other systems

Although the Na–K ATP-ase is the best characterized ATP-dependent transport system, there are a number of other examples where ATP cleavage at a membrane provides the energy for transport. The details of these other systems may however show many differences.

Calcium transport appears to be ATP-dependent in a number of cells. For example, in red blood cells there is an active extrusion of calcium which maintains the intracellular concentration of this ion at very low levels; at higher concentrations calcium inhibits the Na–K ATP-ase and potassium permeability increases. The calcium pump is independent of the sodium pump (the associated ATP-ase being stimulated by internal but not external calcium), and insensitive to sodium, potassium and ouabain.

Another example is provided by the sarcoplasmic reticulum of muscle. This is equivalent to the endoplasmic reticulum and consists of an extensive system of closed lamellae in which calcium is concentrated to a considerable degree; it is important in the regulation of the contractile process of muscle. An ATP-ase which is dependent on magnesium and stimulated by calcium is involved in this transport of calcium.

There is also good evidence that certain transport systems in bacteria and fungi are dependent on ATP. Working with the fungus *Neurospora crassa*, Slayman, Long and Lu (1973) studied the effect of respiratory inhibitors on the resting membrane potential. They observed that the time course of ATP decay coincided with the loss of membrane potential, whereas inhibition of electron transport occurred much more quickly; this suggested that an electrogenic ion pump in the plasma membrane is fuelled by ATP.

Evidence for ATP-driven ion transport in plants is, in general, more equivocal, although there are a number of interesting cases of a more clear-cut relationship. For example, in giant algal cells, cation transport appears to be linked to ATP generated in photosynthesis; the evidence for this is discussed in some detail in one of the case studies in Chapter 5. Another interesting transport system is the salt gland of *Limonium* studied by Hill and Hill (1973). The massive excretion of sodium chloride by the salt glands of various higher plants is a very effective adaptation to saline environments. They are able to regulate their ion content by secreting excess ions. In *Limonium* the secretory system seems to be highly specific for the active transport of chloride ions; when grown in a chloride-free medium, activity of the salt gland ceases. When plants grown in such a medium are

(a)

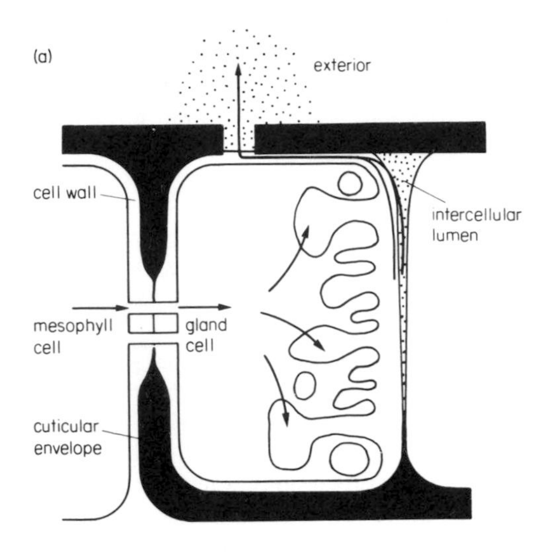

(b)

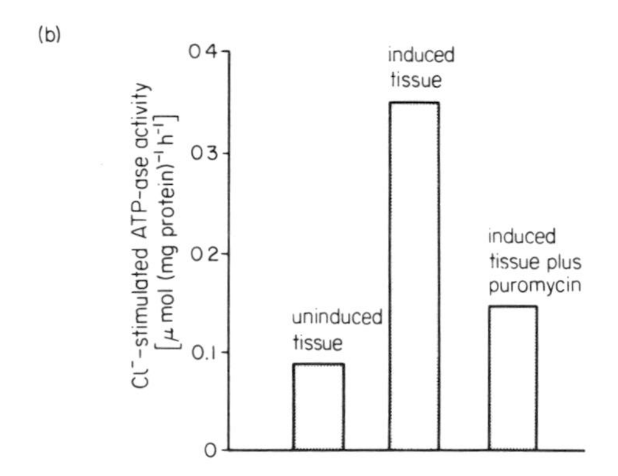

Fig. 4.20 ATP-ase activity in *Limonium* salt gland. (*a*) Diagram of salt extrusion in the gland. Ions enter the gland cell through plasmodesmata from the surrounding mesophyll cells and are pumped into the plasmalemma invaginations of the gland cell. The invaginations and intercellular lumen act as a coupling space for salt and water and the solution leaves a cuticle pore to the exterior. (*b*) Effect of puromycin on the induction of Cl^--ATP-ase activity of the microsomal fraction from *Limonium* leaf tissue. (Redrawn from Hill and Hill, 1973 a & b.)

transferred to one containing chloride, salt excretion increases and reaches a maximum within 3 hours (Fig. 4.20). This adaptation to salinity appears to be associated with an inducible, chloride-stimulated ATP-ase activity located in the microsomal fraction. The development of both chloride

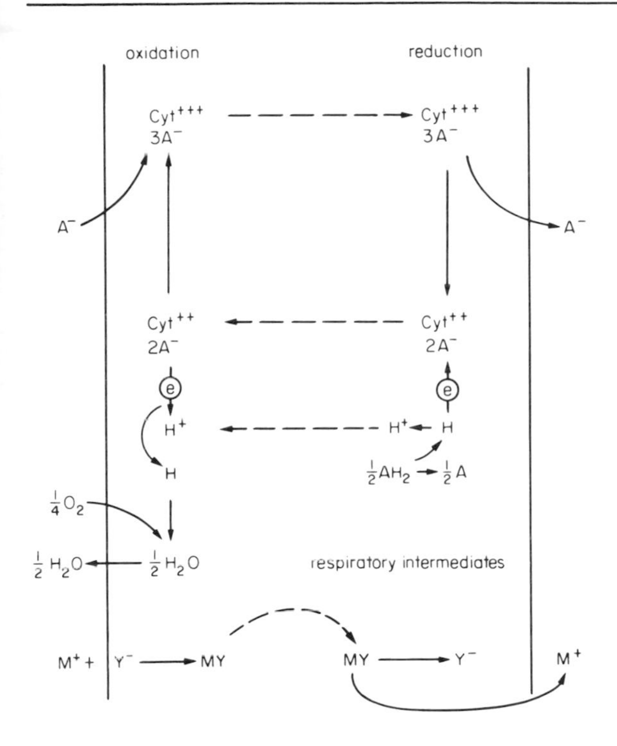

Fig. 4.21 Diagram summarizing Lundegårdh's concept of ion absorption.

excretion and the ATP-ase show a similar sensitivity to the presence of inhibitors of protein synthesis such as puromycin.

Transport linked to electron flow

We have seen that in a number of transporting systems there is good evidence that active ion transport is directly linked to ATP hydrolysis, the ATP being largely produced as a result of electron transport in the mitochondria and chloroplasts. However there are other transport systems, found particularly in mitochondria and in bacteria, which do not depend on ATP or on other high-energy compounds. In these cases active transport appears to be linked more directly to electron flow. This concept of linkage to electron flow has been under consideration for some time, originating with the work of Lundegårdh in the 1930s. He showed that the uptake of both salt and oxygen by plant roots was inhibited by cyanide, indicating a

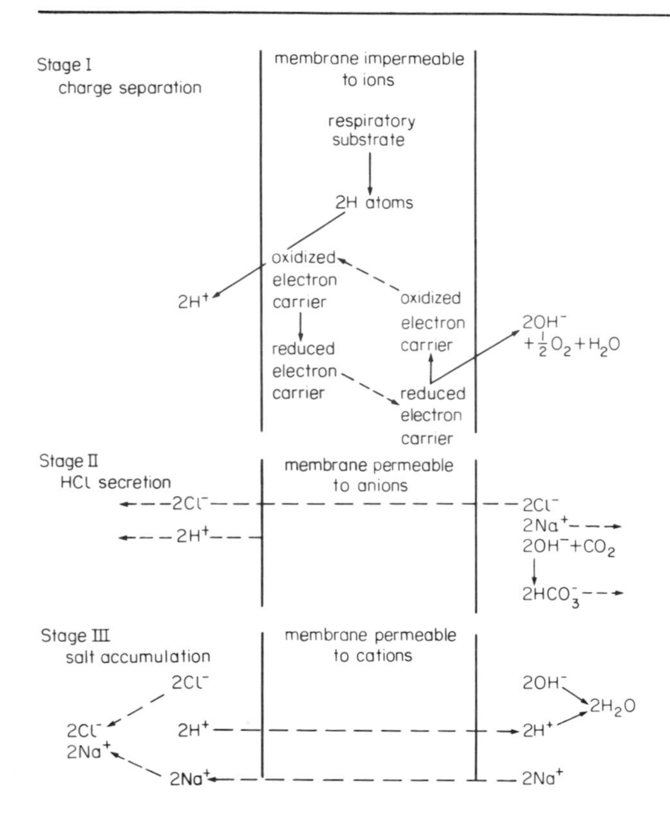

Fig. 4.22 Stage I: the principle of separation of protons and electrons in a membrane due to the orientation of the electron carrier. Stage II: chloride movement from right to left (e.g. blood side to secretory side in gastric mucosa) with an anion permeable stage in the membrane. Stage III: sodium chloride accumulation on the left (e.g. in a plant cell) if the membrane were permeable to hydrogen ions and cations after chloride had moved; water formed on the right. Solid arrows, chemical reactions; broken arrows, diffusion pathways. (Reproduced from Robertson, 1968.)

possible association with the cytochrome system in the respiratory chain, since the terminal step, cytochrome oxidase, is very sensitive to cyanide. Furthermore, when salt was introduced to a plant tissue respiring in water there was an increase in the respiration rate (known as salt respiration) which was quantitatively related to the amount of anion absorbed. Lundegårdh proposed that respiration might be linked to the transport of anions through the cytochrome system, the cytochromes transporting anions inwards as electrons are moved outwards. The cations were thought to move passively down the gradient of electropotential produced by the movement of the anion (Fig. 4.21). Thus the basic feature of this hypothesis is that charge separation associated with electron flow in respiration can provide the driving force for ion transport.

Although this hypothesis is no longer tenable in its entirety, the basic concept linking charge separation and ion transport is considered an essential feature of many transport processes, particularly those associated with organelles. This idea was expanded and modified by Robertson and others (see Robertson, 1968) to connect proton and electron separation with processes such as ion accumulation in plants and proton production in the gastric mucosa (Fig. 4.22). In this scheme the cytochromes are not specifically involved but replaced by some unspecified electron carrier. The membrane involved in this process was not identified, and this highlights what is perhaps the most serious drawback to this concept: it is difficult to see how electron transport which occurs in the mitochondria and chloroplasts is linked to ion transport at the plasma membrane of plant and animal cells. Nonetheless, some evidence has been obtained which is consistent with such a linkage, particularly in relation to anion transport in giant algal cells, and this is discussed further in Chapter 5.

Transport in bacteria

Some direct evidence for a linkage between active transport and electron flow has also come from studies on bacteria and, in particular, work on bacterial membrane vesicles by Kaback and associates (see Kaback, 1973, 1974). These vesicles are prepared by the osmotic bursting of bacterial cells which have been treated with enzymes to degrade the cell wall. These vesicles lack the endogenous metabolism of the intact cell but accumulate a variety of solutes when supplied with an exogenous energy source. Thus they provide an excellent system for the study of the energy source of active transport and of the reactions involved at the membrane. This is much more difficult to discern with intact cells, where added energy sources have little effect on transport unless the cells are severely depleted of reserves. Even then, control and interpretation of experiments are much more difficult due to interference by the endogenous metabolism of the cell.

What then is the energy source for transport in these isolated vesicles? In vesicles prepared from *Escherichia coli* the uptake of some sugars is by the group translocation mechanism discussed earlier (p. 75). In contrast, the uptake of various other sugars, sugar phosphates, amino acids, keto- and di-carboxylic acids is driven by the oxidation of exogenous D-lactate by a membrane bound, flavin-linked D-lactate dehydrogenase. The probable sequence of the respiratory chain in the bacterial membrane is shown in Fig. 4.22. The fall in potential between the primary dehydrogenase and cytochrome b_1 is believed to provide the energy for active transport. Neither the hydrolysis nor synthesis of ATP is involved in this active transport, although other electron donors such as succinate, L-lactate or NADH

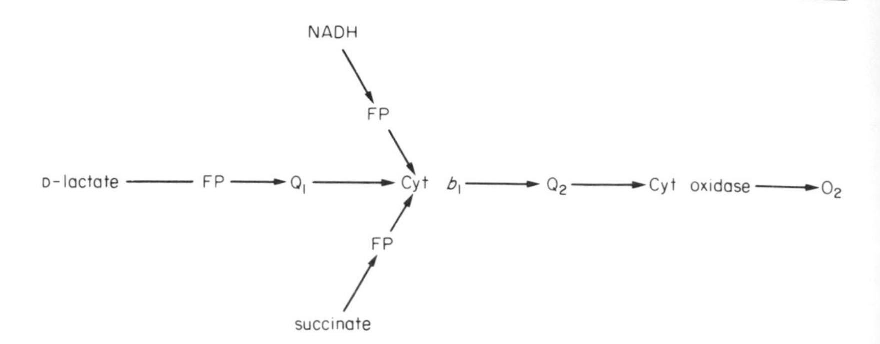

Fig. 4.23 Probable arrangement of components in the respiratory chain of the bacterium *E. coli*; FP, flavoprotein; Q, ubiquinone; Cyt, cytochrome.

may substitute for D-lactate, albeit less effectively. In relation to active inorganic ion transport, these isolated vesicles are found to have lost the ability to transport potassium and rubidium, although interestingly this activity may be restored by addition of the ionophore valinomycin (see p. 66). This valinomycin-induced transport can then be driven by an external energy source, the most effective being D-lactate. In vesicles isolated from other bacteria, transport is driven by the oxidation of other substrates such as L-α-glycerol phosphate, while non-physiological electron donors, such as reduced phenazine methosulphate, can also drive a variety of transport systems.

It is not yet clear how the oxidation of these substrates drives active transport. It was initially thought that electron flow was directly linked to ion transport. Successive oxidation and reduction of a carrier which occupies a site in the electron transport chain between D-lactate dehydrogenase and cytochrome b_1 result in cyclic conformational changes which move the substrate of the carrier across the membrane. However, more recent experiments have raised doubts about this interpretation. For example, mutants lacking D-lactate dehydrogenase have been found which are able to accumulate amino acids to a normal level while, in other cases, transport has been found to be sensitive to uncoupling agents. Significantly, it has also been observed that ATP can drive transport in isolated vesicles under certain conditions (see Simoni and Postma, 1975).

Chemiosmotic coupling

In many respects, this aspect of bacterial transport is similar to transport in mitochondria where phosphorylation, oxidation and transport are all closely linked. Probably the best-supported hypothesis concerning this process is the chemiosmotic theory of Mitchell (see Mitchell, 1970), which has been discussed earlier in relation to oxidative and photosynthetic

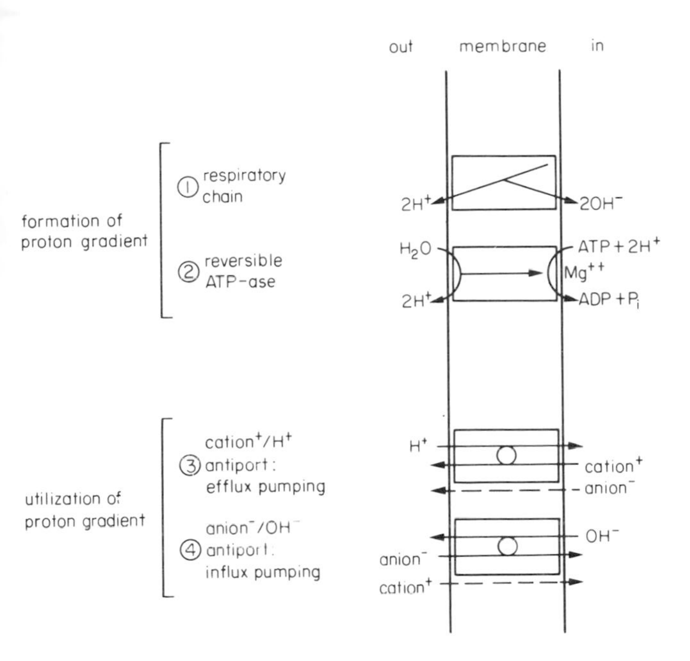

Fig. 4.24 Chemiosmotic hypothesis for the formation of proton gradients across the inner mitochondrial membrane, and the utilization of the resulting proton-motive force in ion transport. The coupled movement of two ions of the same charge in opposite directions in this way is termed 'antiport'. Movement of two ions of opposite charge in the same direction is known as 'symport'. (Redrawn from Hanson and Koeppe, 1975.)

phosphorylation (see p. 78). In this theory it is proposed that the oxidation of electron donors and the associated electron transport chain leads to an extrusion of protons on one side of a membrane which is impermeable to protons and most ions. This generates a *proton-motive force* (PMF) made up of an electric potential difference and a pH difference across the membrane. This PMF may then be used to drive ATP synthesis or transport. These are assumed to be reversible processes, so that downhill solute movement or ATP hydrolysis could generate a proton gradient. Such a scheme for the inner mitochondrial membrane is shown in Fig. 4.24 and may be considered as electrogenic proton pumping. Ion transport is postulated to be mediated by exchange carriers or *'porters'* which exchange protons or hydroxyl ions for cations and anions. Since this exchange is neutral, the electric potential is unaltered and a counter-ion can move down the electrical gradient. Presumably such a PMF could be involved in the active transport of certain solutes in bacterial cells, suggesting that in the isolated

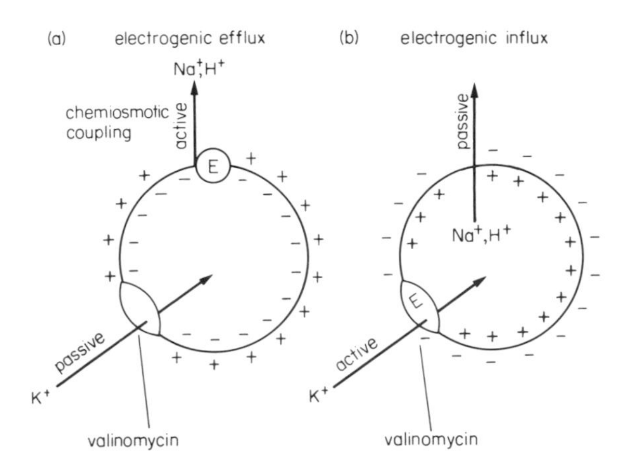

Fig. 4.25 Diagrammatic representation of possible mechanisms for valinomycin-induced potassium (or rubidium) transport in isolated bacterial membrane vesicles. (Redrawn from Kaback, 1973.)

vesicles referred to earlier (p. 97) the ability to generate or utilize ATP is lost. An observation consistent with the above scheme is that lactate oxidation generates a membrane potential that is positive on the outside, but conclusive evidence is still required to establish chemiosmotic coupling in bacterial transport systems. For example, chemiosmosis may be involved in the valinomycin-induced uptake of potassium by isolated vesicles, which is discussed above. Kaback (1973) points out that the active efflux of protons or sodium ions could result in a transmembrane potential which is positive outside and negative inside; valinomycin would facilitate the passive influx of potassium to neutralize the electrical charge across the membrane (Fig. 4.25*a*). An alternative explanation is that valinomycin induces active potassium uptake, resulting in a transmembrane potential which is positive inside, causing a passive efflux of protons or sodium with the electrical gradient (Fig. 4.25*b*). Kaback (1973) considered that the available evidence at that time favoured the latter explanation.

Co-transport

Co-transport is the third distinct mechanism which can supply energy to an active transport process, although this is done indirectly by coupling the uphill transport of one solute to the flow of a second solute down its electrochemical energy gradient. The best-described example is the sodium co-transport of amino acids and sugars in animal cells. The intracellular concentration of sodium is kept low by the Na–K$^+$ ATP-ase pump, which thus maintains a driving force for the entry of sodium. Sodium and the co-transported substrate are moved simultaneously by a facilitated diffusion carrier which requires binding of sodium and the substrate for efficient transport (Fig. 5.5). The accumulation of the substrate is thus dependent on the sodium extrusion pump which requires ATP, although the only link to the carrier is the sodium gradient itself. This sodium-linked co-transport system in mammalian cells is discussed in more detail in Chapter 5.

In other cells (plant, fungal, bacterial) sodium-dependent co-transport of amino acids and sugars is not common. Presumably the sodium co-transport system is an evolutionary development: the cells that lack this mechanism are either less responsive to sodium or do not exist in an environment with a high sodium concentration. In bacteria and yeast, there is evidence that protons can take the place of sodium (see Slayman, 1974), which suggests that co-transport with protons may be more fundamental than co-transport with sodium. For example, in *Escherichia coli*, West and Mitchell (1973) showed that the influx of lactose is accompanied by an influx of protons with a stoichiometry of about 1:1; similar results have been obtained with other bacteria for the uptake of various amino acids. Many of the transport systems of bacteria and mitochondria, which are driven by oxidative reactions, might be considered as examples of co-transport where protons provide the driving force, but this has yet to be fully established. However, consider the uptake of anions by mitochondria (see Fig. 4.24). Phosphate uptake, for example, is probably mediated by an hydroxyl/phosphate counter-transport mechanism which is electrically neutral and is, in effect, equivalent to a proton/phosphate co-transport system.

Energy conservation

We have described evidence that suggests that transport may be driven either by ATP hydrolysis or by electron flow. It is believed that either of these processes can give rise to an as yet unidentified energized state of the membrane, which in turn may be used to drive the active transport of a variety of solutes. This concept is discussed in full by Boyer and Klein (1972) and Christensen (1975) and is summarized in Fig. 4.26. A given active transport process may be driven by the obligatory energized state from energy flowing from either electron transport or ATP hydrolysis, or from another form of energy storage (e.g. a sodium gradient across the membrane). The plasma membranes of higher organisms do not contain an energy-conserving electron transport mechanism, and so the electron transport process should be deleted from the scheme. On the other hand, bacterial membrane vesicles lack the ability to utilize ATP, and so this section should be deleted from the scheme. It is also possible that one of the

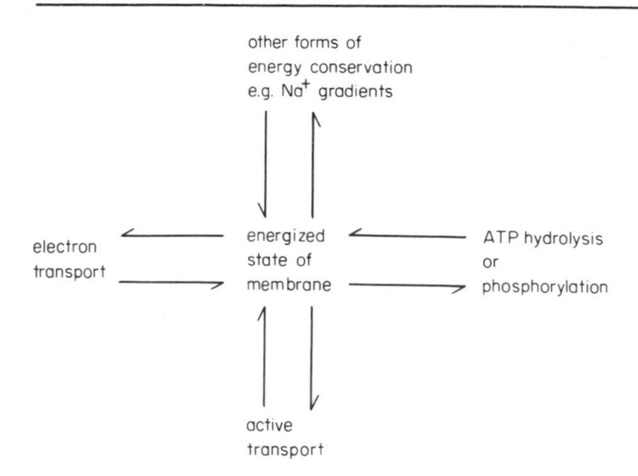

Fig. 4.26 Energy flow to active transport through a common energized state of the membrane. (After Christensen, 1975.)

gradients mentioned, particularly the gradient of electrochemical potential for protons, may represent the energized state itself, rather than a secondary product. Thus the energy storage state between electron transport and phosphorylation in the mitochondrion or chloroplast is perhaps the same as that between electron transport or ATP hydrolysis and solute transport in the plasma membrane.

Energy use in transport

How the energy conserved by the membrane is used to drive solute transport is not understood. Apart from the binding proteins isolated from bacteria (see p. 63) little is known as to the nature of specific membrane carriers. We do not know if the energy supply operates through the carrier or some other membrane component, or at what stage in the transport process is it effective. Clearly much work needs to be done in this area of transport.

A number of models have been proposed to explain energy utilization. One possibility is that energy is used to decrease the affinity of the carriers for the solute, so stimulating solute release at the inner face of the membrane. A more specific model has been proposed by Boyer and Klein (1972). In this, the affinity of the carrier remains unchanged and energy is used to close an inner membrane recognition site after it has opened in response to a loaded carrier and the solute has been discharged (Fig. 4.27). This would allow for a common energization mechanism for a variety of

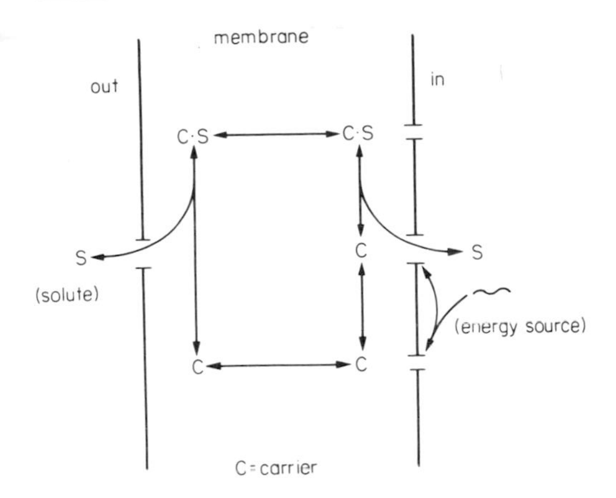

Fig. 4.27 Scheme for carrier-coupled conformational transport. (Redrawn from Boyer and Klein, 1972.)

protein carriers. A simple competition between the rate of combination of solute with carrier and the rate of energy input would determine whether a given solute molecule is retained.

Further reading and references

Further reading

CHRISTENSEN, H. N. (1975) *Biological Transport.* Benjamin, Reading, Mass.

CLARKSON, D. (1974) *Ion Transport and Cell Structure in Plants.* McGraw-Hill, Maidenhead and New York.

DAHL, J. L. and HOKIN, L. E. (1974) The sodium–potassium adenosinetriphosphatase, *Ann. Rev. Biochem.*, **43**, 337.

EPSTEIN, W. (1975) Membrane transport, in *Biochemistry of Cell Walls and Membranes,* Vol. 2, p. 249, ed. C. F. Fox. MTP International Review of Science, Butterworths, London.

HALL, J. L., FLOWERS, T. J. and ROBERTS, R. M. (1974) *Plant Cell Structure and Metabolism.* Longman, London.

KABACK, H. R. (1974) Transport studies in bacterial membrane vesicles, *Science,* **186**, 882.

ROBERTSON, R. N. (1968) *Protons, Electrons, Phosphorylation and Active Transport.* Cambridge University Press, Cambridge, UK.

SIMONI, R. D. and POSTMA, P. W. (1975) The energetics of bacterial active transport, *Ann. Rev. Biochem.*, **44**, 523.

WHITTAM, R. (1975) Kinetic and enzymic aspects of membrane transport, in *Biological Membranes*, Ch. 10, p. 145, ed. D. S. Parsons. Clarendon Press, Oxford.
WHITTAM, R. (1975) Enzymic aspects of sodium pumping across membranes, in *Biological Membranes*, Ch. 11, p. 158, ed. D. S. Parsons. Clarendon Press, Oxford.

Other references

BONTING, S. L. (1970) Sodium–potassium activated adenosinetriphosphatase and cation transport, in *Membranes and Ion Transport*, p. 257, ed. E. E. Bittar. Wiley-Interscience, New York and London.
BOYER, P. D. and KLEIN, W. L. (1972) Energy coupling mechanisms in transport, in *Membrane Molecular Biology*, p. 323, eds C. F. Fox and A. Keith. Sinauer Assoc. Inc., Stamford, Conn.
CALDWELL, P. C., HODGKIN, A. L., KEYNES, R. D. and SHAW, T. I. (1960) The effects of injecting 'energy-rich' phosphate compounds on the active transport of ions in the giant axons of *Loligo*, *J. Physiol.*, **152**, 566.
EPSTEIN, E. (1972) *Mineral Nutrition of Plants: Principles and Perspectives.* Wiley, New York and London.
GREVILLE, G. D. (1969) A scrutiny of Mitchell's chemiosmotic hypothesis of respiratory chain and photosynthetic phosphorylation, in *Current Topics in Bioenergetics*, Vol. 3, p. 1, ed. D. R. Sanadi. Academic Press, London and New York.
HANSON, J. B. and KOEPPE, D. E. (1975) Mitochondria, in *Ion Transport in Plant Cells and Tissues*, p. 79, eds D. A. Baker and J. L. Hall. North-Holland, Amsterdam and London.
HILL, A. E. and HILL, B. S. (1973a) The electrogenic chloride pump of the *Limonium* salt gland, *J. Membrane Biol.*, **12**, 129.
HILL, B. S. and HILL, A. E. (1973b) ATP-driven chloride pumping and ATP-ase activity in the *Limonium* salt gland, *J. Membrane Biol.*, **12**, 145.
HOAGLAND, D. R. (1944) *Lectures on the Inorganic Nutrition of Plants.* Chronica Botanica, Waltham, Mass.
JAGENDORF, A. T. and URIBE, E. (1967) Photophosphorylation and the chemiosmotic hypothesis, *Brookhaven Symp. Biol.*, **19**, 215.
KABACK, H. R. (1973) Active transport in bacterial cytoplasmic membrane vesicles, *Symp. Soc. Expl. Biol.*, **27**, 145.
MITCHELL, P. (1966) Chemiosmotic coupling in oxidative and photosynthetic phosphorylation, *Biol. Rev.*, **41**, 445.
MITCHELL, P. (1970) Reversible coupling between transport and chemical reactions, in *Membranes and Ion Transport*, Vol. 1, p. 192, ed. E. E. Bittar. Wiley-Interscience, New York and London.
SKOU, J. C. (1957) The influence of some cations on an ATP-ase from peripheral nerves, *Biochim. Biophys. Acta*, **23**, 394.
SKOU, J. C. (1960) Further investigations on a Mg^{++} + Na^{+}-activated adenosinetriphosphatase, possibly related to the active, linked transport of Na^{+} and K^{+} across the nerve membrane, *Biochim. Biophys. Acta*, **42**, 6.
SLAYMAN, C. L., LONG, W. S. and LU, C. Y-H. (1973) The relationship between ATP and an electrogenic pump in the plasma membrane of *Neurospora crassa*, *J. Membrane Biol.*, **14**, 305.
SLAYMAN, C. L. (1974) Proton pumping and generalized energetics of transport: a review, in *Membrane Transport in Plants*, p. 107, eds U. Zimmermann and J. Dainty. Springer Verlag, Berlin.
WEST, I. C. and MITCHELL, P. (1973) Stoichiometry of lactrose–protein symport across the plasma membrane of *Escherichia coli*, *Biochem. J.*, **132**, 587.
WHITTAM, R. (1958) Potassium movements and ATP in human red cells, *J. Physiol.*, **140**, 479.

WHITTAM, R. and AGER, M. E. (1965) The connection between active cation transport and metabolism in erythrocytes, *Biochem. J.*, **97**, 214.
WHITTAM, R. and BLOND, D. M. (1964) Respiratory control by an adenosine triphorphatase involved in active transport in brain cortex, *Biochem. J.*, **92**, 147.
WHITTAM, R. and CHIPPERFIELD, A. R. (1975) The reaction mechanism of the sodium pump, *Biochim. Biophys. Acta*, **415**, 149.

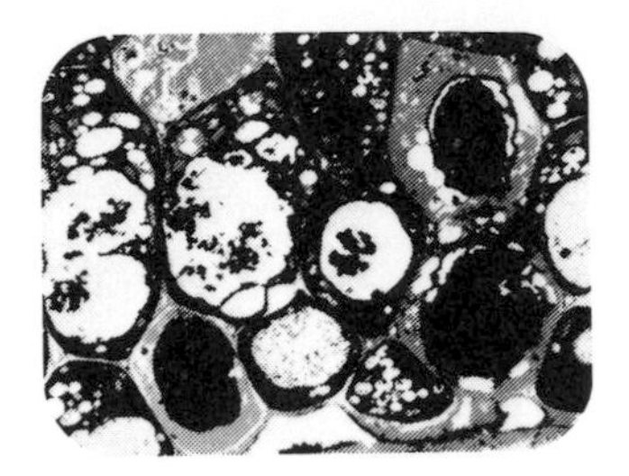

Chapter 5

Case studies of specific transport mechanisms

In this chapter we will consider two extensively studied transport systems in order that we may illustrate the application of some of the principles and concepts outlined in the preceding chapters. For this purpose we will present results obtained with two very different experimental materials. At the cell level, giant freshwater algae have provided an experimental system in which ion transport has been partially characterized in a series of extensive and detailed investigations. These studies form our first case study. Certain studies on animal systems have been concentrated on the polar transport of solutes across epithelial cell layers in which a tight junction exists between the cells, thus preventing extracellular passage of solutes across the layer. In particular the co-transport of organic solutes such as sugars and amino acids in association with sodium has been the subject of intensive investigation, and will be presented as our other case study.

Ion transport in giant algal cells

Extensive ion flux measurements have been made on four freshwater species of giant alga, *Nitella translucens, Chara corallina, Tolypella intricata* and *Hydrodictyon africanum.* The first three are Characean species, the overall structure of which is illustrated in Fig. 5.1. The large vacuolate internodal cells may be up to 100 mm in length, and 800–1,000 μm in diameter. The cytoplasm is in two layers, the outer 5–6 μm is a stationary gel layer in which the chloroplasts are embedded in a regular array, the inner 5–6 μm is a streaming layer free of chloroplasts which moves at about 50 $\mu m\, s^{-1}$ parallel to the stationary gel layer. *Hydrodictyon africanum* belongs to the Chlorococcales and consists of giant spherical cells 3–6 mm in diameter derived from colonial nets of 256 or 512 cells of identical genetic composition. Individual cells may increase to more than 10 mm in diameter. The large central vacuole is surrounded by a thin layer of cyto-

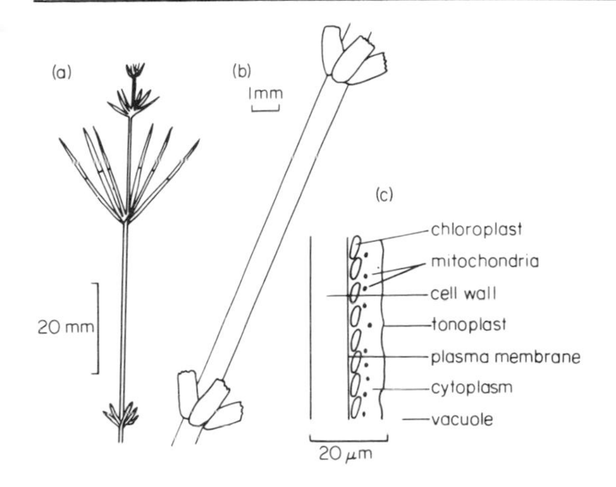

Fig. 5.1 Structure of Charophytes: (*a*) Apex of plant showing elongate internodal cells. Growth is apical. (*b*) A single internodal cell. (*c*) Longitudinal section of an internodal cell. (After Briggs *et al.*, 1961.)

plasm which is partially filled by a large complex irregular chloroplast.

Of the above four species, *Nitella* and *Hydrodictyon* are basically similar in relation to their ion-transporting properties, and we will concentrate on the results obtained with these two species.

The electric potentials and ion concentrations of sodium, potassium and chloride in the vacuole and cytoplasm of *Nitella* have been measured (Table 5.1), enabling the Nernst potential E_{Nj} and the electrochemical potential difference ΔE_j to be calculated from Eq. [2.6] using the relationship

$$\Delta E_j = E_M - E_{Nj}. \qquad [5.1]$$

The measured electric potential E_M across the plasma membrane is -138 mV (cytoplasm negative), and across the tonoplast +18 mV (vacuole positive) giving a resultant potential between the vacuole and the outside solution of $(-138 + 18) = -120$ mV. The observed pattern is very similar in *Hydrodictyon* (Raven, 1967). It is apparent from the data presented in Table 5.1 that none of the calculated Nernst potentials match the measured membrane potential across the plasma membrane. For all three ion species, the electrochemical potential difference between the cytoplasm and the outside ΔE_j^{co}, has a large value. Thus we can state that sodium is at a lower electrochemical potential in the cytoplasm than outside and therefore will tend to move passively inwards; potassium and chloride are at a higher electrochemical potential in the cytoplasm than outside and will tend to move passively outwards. Across the tonoplast the situation is different,

Table 5.1 Calculation of Nernst potentials E_{N_j} from the measured concentrations of sodium, potassium and chloride in the cytoplasm and vacuole of *Nitella translucens* using the measured electrical potential profile. The suffixes *o, c, v* refer to outside solution, cytoplasm and vacuole, respectively.

Ion	c_j^o (mol m^{-3})		$E_{N_j}^{co}$ (mV)	c_j^c (mol m^{-3})	$E_{N_j}^{cv}$ (mV)		c_j^v (mol m^{-3})	Ion
Na^+	1.0		− 66	14	−39		65	Na^+
K^+	0.1		−178	119	−12		75	K^+
Cl^-	1.3		+ 99	65	−23		160	Cl^-
	Outside	Plasma membrane		Cytoplasm		Tonoplast	Vacuole	

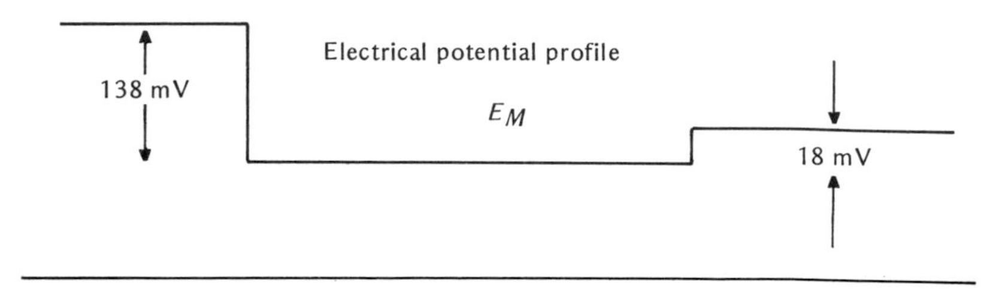

(From Spanswick and Williams, 1964.)

the sodium being at a lower electrochemical potential in the cytoplasm than it is in the vacuole, and thus it will tend to move passively from vacuole to cytoplasm; potassium and chloride are close to passive equilibrium across the tonoplast.

This situation would indicate that if ions in *Nitella* move passively the influx of sodium across the plasma membrane should exceed the efflux and the effluxes of potassium and chloride should exceed the influx. It is possible, using the Ussing–Teorell flux ratio equation, [2.7], to calculate the actual ratio of such passive fluxes. From the data of Table 5.1 $\phi_{Na}^{oi}/\phi_{Na}^{io}$ is approximately 12/1, ϕ_K^{oi}/ϕ_K^{io} is approximately 1/5 and $\phi_{Cl}^{oi}/\phi_{Cl}^{io}$ is approximately 1/10,000. Similar ratios have been reported for *Hydrodictyon* (Raven, 1967). Plasma membrane fluxes have been measured in *Nitella* (MacRobbie, 1962, 1964) for sodium, potassium and chloride and found to be between 5 and 10 nmol m^{-2} s^{-1} for these ions (see Table 5.2). For each of these ions the influx and efflux are nearly equal. Thus the measured fluxes do not agree with the predicted passive pattern and work must therefore be done to oppose the passive movement of these ions across the plasma membrane.

The magnitude of the work involved can be calculated from the value

Table 5.2 Fluxes of sodium, potassium and chloride across the plasma membrane of *Nitella translucens* measured in the light and in the dark.

	Light		Dark	
	ϕ_j^{oi}	ϕ_j^{io}	ϕ_j^{oi}	ϕ_j^{io}
Ion	**(nmol m^{-2} s^{-1})**			
Na^+	5.5	5.5	5.5	1.0
K^+	8.5	8.5	2.0	(8.5)
Cl^-	8.5	8.5	0.5	—

(From MacRobbie, 1962, 1964.)

of ΔE_j. As 1 mV equals 23.06 cal mol^{-1}, for sodium the energy required is 1,660 cal mol^{-1}, for potassium 922 cal mol^{-1} and for chloride 5,465 cal mol^{-1}. This energy must be supplied by the metabolism of the cell, and the transport of these ions in the opposite direction to the predicted passive fluxes will therefore be active. The sodium will be actively pumped outwards at the plasma membrane, while potassium and chloride will be actively pumped inwards.

The source of this energy in green plant cells may be either photosynthetic or respiratory (see p. 76). The effect of light on the ion fluxes in *Nitella* is presented in Table 5.2, where it can be seen that sodium efflux,

Table 5.3 Active ion fluxes in illuminated cells of *Hydrodictyon africanum* in the presence and absence of carbon dioxide in air at 14°C.

	Ion fluxes (nmol m^{-2} s^{-1})		
	ϕ_{Na}^{io}	ϕ_K^{oi}	ϕ_{Cl}^{oi}
Normal air	3.7	12.4	20.4
Carbon-dioxide-free air	3.6	10.4	20.9

(From Raven, 1967b.)

potassium influx and chloride influx are reduced in the dark, implicating photosynthesis as the major energy source. It has been demonstrated that carbon dioxide fixation is not a requirement for energizing these pumps, active ion transport continuing in the light in the complete absence of CO_2 (Table 5.3). It would therefore appear that it is the light reactions of photosynthesis in which light energy is absorbed and converted into chemical potential energy which drive active transport. A scheme for these

Table 5.4 The effect of the uncoupler CCCP on active ion fluxes in illuminated cells of *Hydrodictyon africanum*.

	Ion fluxes, as percentage of uninhibited control		
CCCP (mol m^{-3})	ϕ_{Na}^{io}	ϕ_{K}^{oi}	ϕ_{Cl}^{oi}
0	100	100	100
1×10^{-3}	68	76	107
5×10^{-3}	35	66	106
1×10^{-2}	38	45	88

(From Raven, 1967b.)

reactions is presented and discussed in Chapter 4 (Fig. 4.8), from which it can be seen that there are two photochemical reactions referred to as photosystems I and II (PS I and PS II). PS I absorbs light up to a wavelength of 730 nm and PS II, which requires higher energy quanta, absorbs light up to a wavelength of 705 nm.

When *Nitella* is illuminated with red light between 705 and 730 nm only PS I is excited. There is no reduction of NADP and no oxygen production, but ATP is still generated by the cyclic phosphorylation pathway. Under these conditions the active potassium influx and active sodium efflux are unaffected but the active chloride influx is inhibited, suggesting that the cation fluxes are ATP-driven and that the chloride flux is associated with PS II.

When both photosystems are operative in white light the cation fluxes can be inhibited with uncoupling agents, such as carbonylcyanide *m*-chlorophenylhydrazone (CCCP) or imidazole, which prevent ATP formation, while the active chloride influx is unaffected or even slightly stimulated (Table 5.4). The observed stimulation may be a reflection of the increased rate of electron transport which results from the uncoupling of ATP synthesis, if the chloride influx is dependent on the electron transport process. An inhibitor of PS II, dichlorophenyldimethyl urea (DCMU), inhibits the active chloride efflux without affecting the active cation fluxes. These experimental results suggest that the cation pumps are ATP-energized but that the chloride pump may be closely linked with the electron transport system of PS II, or is dependent on a product other than ATP, such as some reduced substrate.

The movement of other anions has been less extensively studied in *Nitella*. There is evidence for an active transport of phosphate, sulphate and nitrate at sites somewhere between the medium and the vacuole in many large algae, but at present the mechanism is unknown. This ignorance of the transport mechanism for the major essential anions within the plant cell is a serious gap in our current knowledge.

It is well known that a coupling often exists between potassium and

Table 5.5 The effect of the cardiac glycoside ouabain on active and passive ion fluxes in illuminated cells of *Hydrodictyon africanum*. Note that the passive potassium efflux is stimulated by the inhibitor.

Ouabain (mol m^{-3})	Ion fluxes, as percentage of uninhibited control		
	ϕ^{io}_{Na}	ϕ^{oi}_{K}	ϕ^{io}_{K}
0	100	100	100
0.5	72	63	128

(From Raven, 1967a.)

sodium fluxes in animal systems (see p. 85), and investigators have sought to find if a similar coupling exists within plant systems. The cardiac glycoside ouabain has been employed as a specific inhibitor of this potassium/sodium pump in animal cell membranes due to the high affinity with which it binds to ATP-ase, blocking the site for ATP on the pumping mechanism (see p. 89). In *Nitella*, ouabain at a concentration of 5 X 10^{-2} mol m^{-3} reduced potassium influx in the light to near its dark value (MacRobbie, 1962). In *Hydrodictyon* both potassium influx and sodium efflux are similarly inhibited by ouabain (Table 5.5), and sodium efflux is dependent on the external concentration of potassium. These data lend support to the view that the potassium/sodium pump at the plasma membrane of these cells is coupled. That portion of the cation fluxes which is insensitive to ouabain is linked with the chloride entry, showing sensitivities to light and inhibitor very similar to those of the chloride transport. This interdependence is however incomplete, only a portion of the chloride entry being associated with potassium and sodium influx, the measured cation influx always being less than the chloride influx.

An alternative to the above primary active transport of chloride coupled to potassium and sodium influx has been proposed by Smith (1970) who suggests an indirect linkage via proton fluxes (Fig. 5.2). He has suggested that an active proton efflux is partly balanced by a cation influx, the remainder of the active proton efflux being balanced by a passive proton influx in secondary active chloride transport. A downhill hydroxyl efflux is postulated to energize the active chloride influx. Supporting evidence for this model is provided by the observation that chloride influx is stimulated when a pH gradient is created across the plasma membrane by changing from a high to a low pH in the external solution, a situation which should favour proton efflux.

An interesting related phenomenon frequently observed in charophytes is an encrustation of the outer face of the cell wall with a deposit of calcium carbonate. In *Nitella* this deposit is in a banded formation and has been demonstrated to be the result of patches of differing pH (Fig. 5.3). It has been proposed that the cell surface is normally acidic, probably due to the

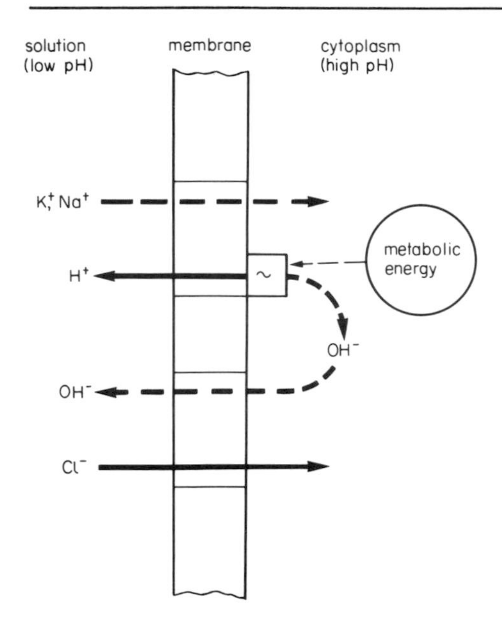

Fig. 5.2 A scheme for net salt transport at the plasma membrane of algal cells. The primary energy-requiring process is proton efflux, which is partly balanced by potassium and sodium influx. The remainder of the proton efflux is balanced by an hydroxyl efflux coupled to an active chloride influx. (After Smith, 1970.)

action of a proton efflux pump. In the presence of bicarbonate, uptake of this anion results in a hydroxyl efflux, the process being regenerative and thus raising the pH in localized areas. The resultant calcium carbonate precipitation encourages further hydroxyl efflux in that area (Fig. 5.4). As a result of the banding phenomenon the chloride influx referred to above is liable to be lower in the alkaline zones than in the acid regions.

In addition to the fluxes discussed in the preceding paragraphs, there are a number of residual ion fluxes in *Nitella* and in *Hydrodictyon* which are insensitive to ouabain and independent of external chloride. These include a portion of the cation influxes and effluxes and the chloride efflux. The residual cation influxes which have been measured in sulphate solutions in the presence of ouabain are light dependent, requiring PS II. If these fluxes are passive, as some investigators have suggested, then the passive cation permeabilities, P_K and P_{Na}, must be lower in the dark. In contrast, chloride efflux, which increases in the dark, would require an increase of P_{Cl} in the dark.

It was suggested above that the main chloride influx pump may require a close link with an electron transport system. However, as such systems

Fig. 5.3 The accumulation of alkali (dark areas) and acid (light areas) outside two isolated internodal cells of *Nitella clavata.* The external medium contained phenol red indicator and mineral salts, including $KHCO_3$; the pH was 6.9. The cells had been in the dark for 20 min and then light (about 2 W m^{-2}) for 20 min before the photograph. (After Spear *et al.*, 1969.)

are restricted to the cell organelles in the cytoplasm, it is difficult to envisage how events in the organelles can be transmitted to the plasma membrane through an appreciable thickness of highly buffered cytoplasm. One possibility is that membrane-bound vesicles are shunted between the organelles and the plasma membrane carrying protons, which might be discharged in exchange for other cations and anions, possibly fuelling the postulated proton extrusion pumps at the plasma membrane.

In the brief outline given above we have presented the results of some recent studies on ion transport in *Nitella* and *Hydrodictyon.* The functions of this transport are diverse and will now be briefly considered. In walled plant cells, a turgor pressure is required to bring about extension growth of the cell wall. Accumulated inorganic ions will contribute to the osmotic

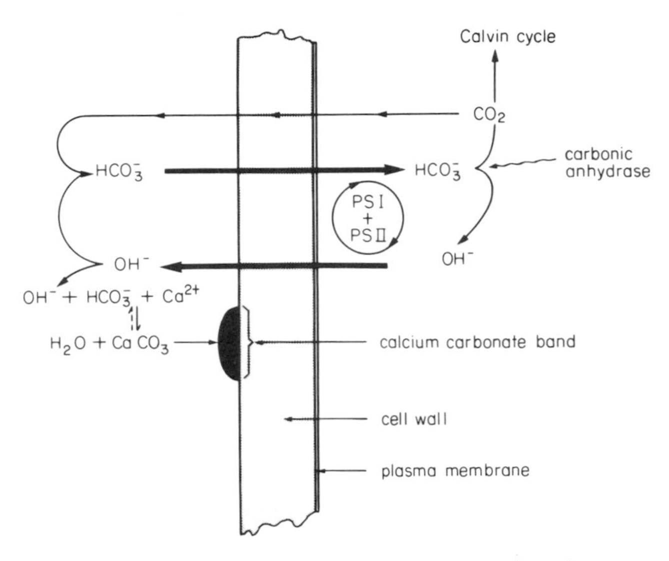

Fig. 5.4 A scheme proposed by Lucas and Smith (1973) for the efflux of hydroxyl ions driven by an active bicarbonate ion influx. Subsequent reaction with calcium and bicarbonate ions leads to the formation of bands of calcium carbonate on Charophyte cells. (After Hope and Walker, 1975.)

pressure and thus increase cell turgor, although it should be noted that organic solutes are also often involved. Similarly, osmoregulation can be achieved by changes in the ion-transporting properties of the organism. The inorganic nutrition of a cell requires not only the accumulation of essential elements but also the removal of toxic solutes such as protons and hydroxyl ions produced during metabolism. Current interest in ion transport in plant cells is now being focused on the mechanisms whereby the fluxes of ions are regulated in relation to the cell's growth and metabolism in terms of both long-term maintenance of the transport systems at a reasonable level and short-term fine control of the processes involved.

Co-transport of organic solutes

In recent years a considerable volume of evidence has accumulated which shows that the transport of sugars and amino acids in a variety of animal cell membranes is dependent upon the presence of sodium. The first report of this phenomenon was the observation that the uptake of glycine or alanine by duck red blood cells was strongly inhibited when part of the sodium in the bathing medium was replaced by potassium (Table 5.6). Further work with Ehrlich ascites tumour cells demonstrated that the same effect was obtained when choline rather than potassium replaced

Table 5.6 The effect of substituting potassium for sodium on the ability of duck red blood cells to accumulate amino acids.

Amino acid	$[K^+]^o$ (mol m^{-3})	$[Na^+]^o$ (mol m^{-3})	Δ [amino acid] in / Δ [amino acid] out
Glycine	0	140	1.69–2.56
	80	60	1.12
	120	20	0.28
	140	0	0.00
Alanine	0	140	0.85–1.10
	105	35	0.71
	140	0	0.12

(From Christensen *et al.*, 1952.)

sodium. Furthermore, during the uptake of glycine or of tryptophan there was a loss of potassium with a concomitant gain of sodium (see Christensen, 1975). It was tentatively postulated at that time that amino acid accumulation was related to potassium movement out of the cell, although the possible relationship with sodium movement into the cell was not excluded. Detailed studies over short time intervals indicated that for a wide range of concentrations, when sodium was replaced by potassium in the extracellular fluid, it was the decreased sodium not the increased potassium which had the largest influence on amino acid uptake. This result was achieved by replacing sodium or potassium with either lithium or choline (see Table 5.7).

Further evidence of sodium involvement in the transport of monosaccharides across the small intestine of toad and rat, the absorption of amino acids by intestine, thymus nuclei, kidney slices, isolated intact dia-

Table 5.7 The effect of various cations on glycine influx into pigeon red blood cells at 39°C.

[KCl]^o (mol m^{-3})	[NaCl]^o (mol m^{-3})	[LiCl]^o (mol m^{-3})	[Choline Cl]^o (mol m^{-3})	Glycine influx (mol m^{-3} 15 min^{-1})
106	40	–	–	0.23
146	–	–	–	0.041
12	40	94	–	0.23
116	30	–	–	0.22
116	–	30	–	0.042
116	–	–	30	0.040
51	95	–	–	0.59

(From Vidaver, 1964.)

phragm and brain slices firmly established the importance of sodium in the transport of these organic solutes.

Sodium-dependent transport is not the only means by which amino acids are transported, neutral amino acids having both sodium-dependent and sodium-independent transport systems. Similarly, the movement of sugars into muscle and red blood cells does not require sodium and appears to take place by facilitated diffusion (p. 62).

A distinction has been made between alanine-type amino acids (A system), which show dependence on sodium, and leucine-type amino acids (L system), whose transport has no significant dependence on sodium. It appears that the length of the non-polar side chain and the resultant degree of polarity of the amino acid are the features which may distinguish these two transport systems. The two systems can function independently of one another and a number of amino acids will react with both systems.

The phenomenon of sodium-stimulated polarized transport of sugars and amino acids into epithelial cells from low concentrations in the mucosal solution to higher concentrations in the serosal solution has been indicated in a number of investigations. That this transepithelial transport is dependent on the presence of sodium in the mucosal solution has been shown by the investigations of Crane and his co-workers, who have demonstrated that sodium is required for the entry of transported sugars into epithelial cells. The proposed mechanism for this transport is presented in Fig. 5.5 in which the proposed mobile carrier has two specific binding sites, one for the organic substrate and one for sodium. Sugar transport has been shown to be a function of the sodium concentration in the medium (Fig. 5.6), potassium and related ions such as lithium and ammonium strongly inhibiting the sugar transport by interfering with sodium carrier interactions.

The absorption of sugars and amino acids by small intestine is accompanied by an increase in the rate of sodium absorption. This has been shown using the short-circuit technique (p. 44) on isolated rabbit ileum. The current which is necessary to abolish the electric potential difference across the membrane, and is therefore a measure of the rate of sodium transport from mucosa to serosa, is increased following the addition of a sugar, 3-*O*-methylglucose, to the solution bathing the mucosal surface (Fig. 5.7). The subsequent addition of phloridzin, which is known to inhibit the sugar carrier, reduces the magnitude of the short-circuit current, reflecting a decreased sodium transport (Schultz and Zalusky, 1964). Further, the increase in short-circuit current is a saturable function of the concentration of sugar or amino acid, as illustrated by the results obtained for L-alanine (Fig. 5.8). The observation that non-metabolized sugars, such as 3-*O*-methylglucose, increase the rate of net sodium transport indicates that the mechanism is independent of the metabolic fate of the solute which is being transported. This reciprocal relation strongly suggests a coupling between sodium transport and sugar or amino acid transport at the brush border of the intestinal cell.

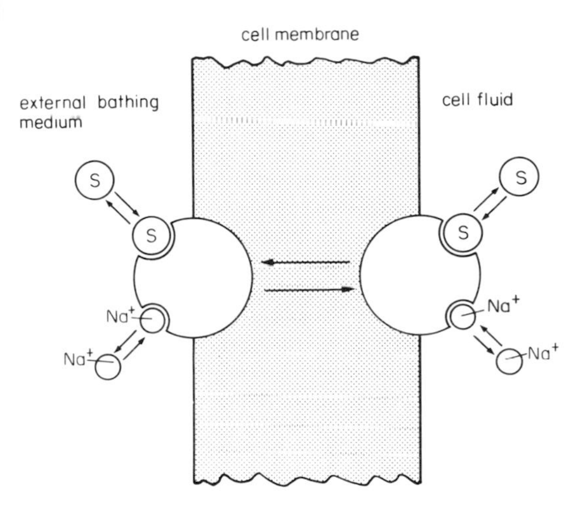

Fig. 5.5 A mobile carrier with specific binding sites for organic solutes S and for sodium.

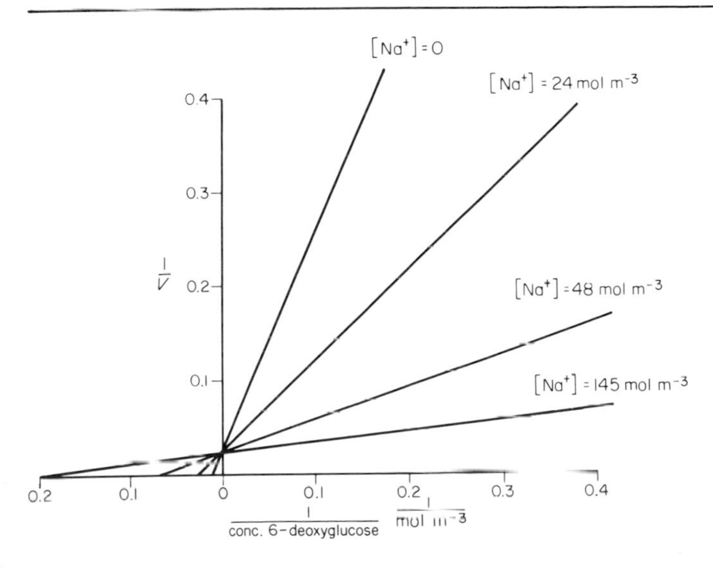

Fig. 5.6 Lineweaver and Burk plot showing the dependence of the rate of 6-deoxyglucose transport on the concentration of 6-deoxyglucose and sodium. (From Crane *et al.*, 1965.) Note that the influx of 6-deoxyglucose conforms to Michaelis–Menten kinetics in both the absence and presence of sodium.

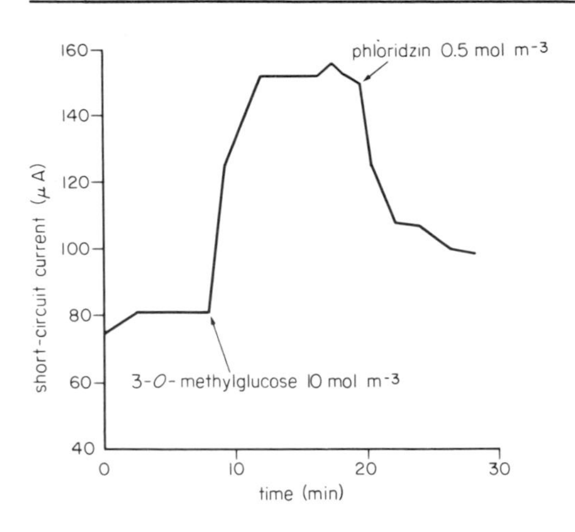

Fig. 5.7 The effect of 3-*O*-methylglucose and phloridzin on the short-circuit current across isolated rabbit ileum. (From Schultz and Zalusky, 1964.)

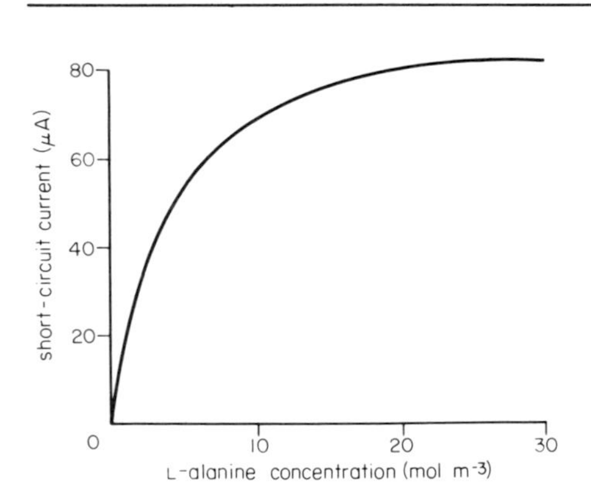

Fig. 5.8 The increase in short-circuit current across isolated rabbit ileum as a function of L-alanine concentration in the mucosal bathing solution. (From Schultz and Zalusky, 1965.)

A model which is consistent with the above observations has been proposed by Crane (1965) and is presented here in Fig. 5.9. Sodium, which is high in the bathing mucosal medium, is postulated to move downhill into the epithelial cell, complexed with a carrier and sugar. Internally, the complex breaks down releasing both sugar and sodium, the sugar thus being accumulated against its concentration gradient by the movement of

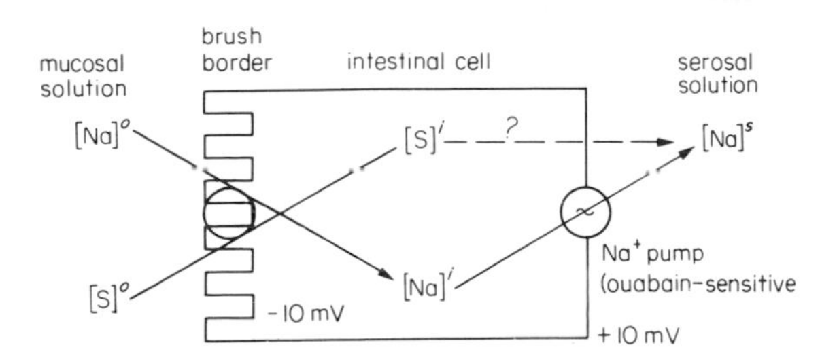

Fig. 5.9 A model for sodium-linked organic solute transport across the small intestine. The E_M value is that reported by Wright (1966).

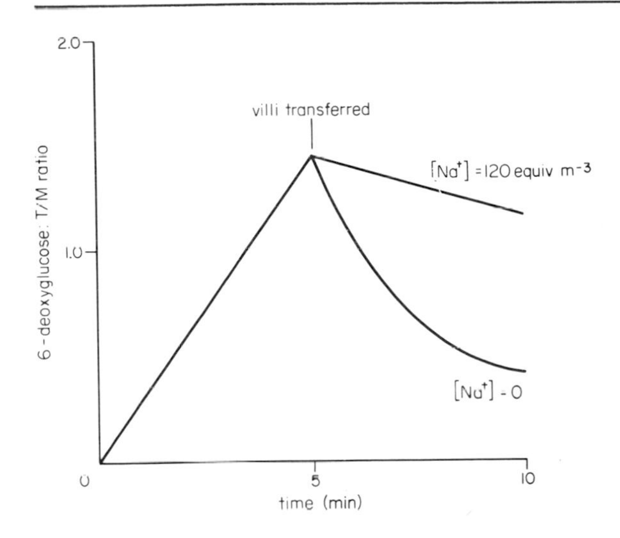

Fig. 5.10 The efflux of 6-deoxyglucose against its concentration gradient (tissue-to-medium ratio, T/M < 1) induced by a reversal of the sodium gradient. (From Crane, 1965.)

sodium down its gradient of electrochemical potential. The gradient for sodium is possibly maintained by a sodium extrusion pump situated at the serosal surface, which would prevent a build-up of sodium within the cell. There is evidence that this pump is ouabain-sensitive, implicating ATP-ase activity (see p. 89).

The sodium-gradient hypothesis is very attractive in that it explains active transport of organic solutes simply in terms of a linkage with sodium. Since a downhill movement of sodium is postulated to result in an uphill movement of sugar, a test of the hypothesis would be to reverse the sodium gradient and observe whether sugar is moved out of the cell into the medium. This has been achieved by Crane (1964) in an experiment presented here in Fig. 5.10. Preparations of isolated villi were incubated under nitrogen in the presence of DNP to limit sodium pumping and concomitant sugar accumulation. After 5 minutes in the presence of 6-deoxyglucose and a high sodium concentration (120 equiv. m^{-3}) the tissue-to-medium ratio of 6-deoxyglucose was slightly above unity, indicating an accumulation of the sugar. The villi were then transferred to a new medium, either identical in composition, or with tris(hydroxymethyl)aminomethane ($Tris^+$) replacing sodium. In the absence of external sodium, 6-deoxyglucose moved out of the cells into the medium against its own concentration gradient, and the tissue-to-medium (T/M) ratio became less than unity.

Another test for the sodium gradient hypothesis is to estimate whether the electrochemical potential difference for sodium across the mucosal membrane is adequate to account for the observed uphill transport of organic solutes. If we consider an organic solute such as alanine moving into the cell coupled to the sodium influx, work will have to be done and there will be a decrease in the total free energy G, such that

$$\frac{dG}{dt} = \phi_A \Delta\bar{\mu}_A + \phi_{Na}\Delta\bar{\mu}_{Na} < 0\,, \tag{5.2}$$

where

$$\Delta\bar{\mu}_A = RT \ln([A]^i/[A]^o) \tag{5.3}$$

and

$$\Delta\bar{\mu}_{Na} = RT \ln([Na]^i/[Na]^o) + FE_M\,, \tag{5.4}$$

ϕ_A and ϕ_{Na} are the coupled net fluxes of alanine and sodium from outside the cell to inside the cell, F is the Faraday, and E_M the membrane potential. The efficiency of the energy conversion is not known, but taking the maximum tissue-to-medium ratio resulting from 100% efficiency (i.e. when $dG/dt = 0$),

$$\left(\frac{[A]^i}{[A]^o}\right)^{\phi_A} = \left(\frac{[Na]^o}{[Na]^i}\right)^{\phi_{Na}} \exp[-\phi_{Na}\, FE_M/RT]\,. \tag{5.5}$$

When $[Na]^o = 140$ mol m^{-3}, $\phi_A = \phi_{Na}$ (Curran *et al.*, 1967) and reasonable values for $[Na]^i$ and E_M are 20–50 mol m^{-3} and 10–20 mV, respectively. Substituting these values into Eq. [5.5], maximum values of the tissue-to-medium ratio for alanine are 4–15. These values are similar to experimentally determined ratios of 8–10 (Schultz *et al.*, 1967), suggesting that the proposed sodium gradient model is energetically feasible, if all the energy available from the downhill movement of sodium is converted to driving the uphill movement of alanine.

Although appealing in its simplicity and supported by a considerable

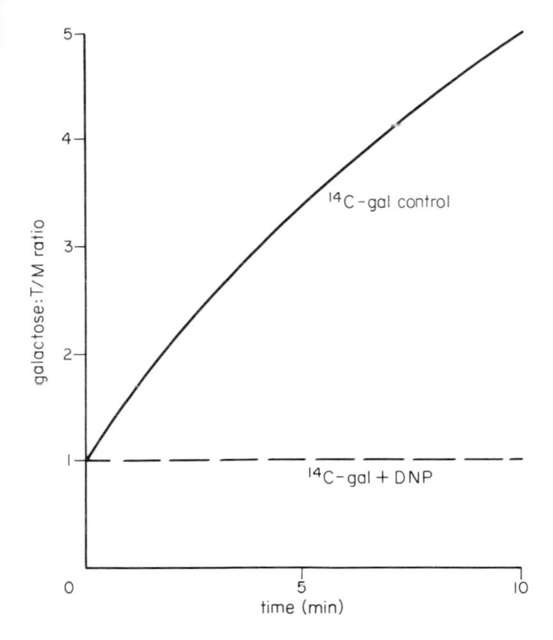

Fig. 5.11 Fluxes of ^{14}C-galactose following the reversal of the sodium gradient in isolated intestinal epithelial cells (T/M = tissue-to-medium). The cells were pre-loaded at 0°C in media with 80 equiv m^{-3} sodium and 1.25 mol m^{-3} ^{14}C-galactose. Incubation at 37°C in 20 equiv m^{-3} sodium. (From Kimmich, 1972.)

volume of experimental evidence, the sodium gradient hypothesis has not been accepted by all investigators. Kimmich (1970) has demonstrated an accumulation of galactose by isolated intestinal epithelial cells against a concentration gradient even when sodium in the cells was at a higher concentration than that in the medium (Fig. 5.11). Furthermore, rapidly transported organic solutes such as valine should inhibit the entry of slowly transported solutes such as 3-*O*-methylglucose due to a partial dissipation of the sodium gradient by the rapid uptake of valine. This inhibition would be dependent on an equal coupling of sodium and valine and of sodium and 3-*O*-methylglucose, which has been established for high external sodium levels ($>$ 100 mol m^{-3}). However, it has been demonstrated that the rapidly transported valine is a much poorer inhibitor than the slowly transported 3-*O*-methylglucose. These and other observations have led Kimmich (1970) to propose that intestinal transport of sugars and amino acids is energized directly by metabolism rather than by the sodium gradient. The proposed alternative scheme for this transport is presented in Fig. 5.12. A membrane-bound ATP-ase is envisaged which provides energy for the transport of

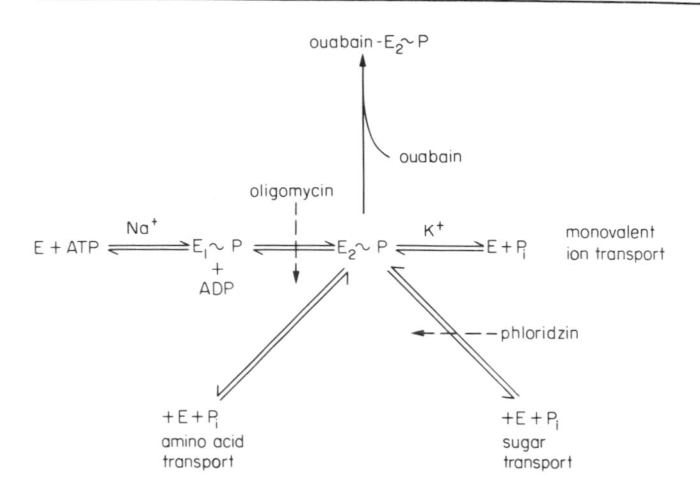

Fig. 5.12 A scheme for the transport of sugars and amino acids energized directly by metabolism. (After Kimmich, 1970.)

sugars and amino acids. Sodium is required by this process for the generation of the required energized intermediate, while potassium is inhibitory because it dissipates energy in monovalent ion transport. Oligomycin is inhibitory because it prevents the formation of the energized intermediate, while ouabain is inhibitory because it prevents the normal utilization of the intermediate as an energy donor. Phloridzin specifically inhibits the transport of sugars.

Additional evidence for and against each of these two hypotheses is provided by the results of other investigations. The uptake of amino acids by Ehrlich ascites tumour cells shows a stoichiometry with an associated sodium uptake (Schafer and Jacquez, 1967). Even when completely inhibited metabolically, Ehrlich cells continue to transport glycine as long as a favourable gradient for sodium exists. Interaction between the movement of amino acids and of sodium has been shown to be mutual; excess extracellular glycine which creates a downhill chemical gradient for amino acid entry also markedly enhances the entry of sodium. These findings obviously favour the co-transport hypothesis but do not necessarily exclude the direct coupling of amino acid transport to metabolism.

Reduction of extracellular sodium levels reduces the influx of amino acids such as glycine, glutamate and amino-isobutyrate without appreciable delay. Changes in cellular sodium concentrations do not by themselves affect amino acid influx, except that internal sodium concentrations greater than 100 mol m^{-3} slightly inhibit it. Also an active transport of amino acids in Ehrlich cells and mouse pancreas cells which is apparently independent of the sodium gradients has been demonstrated (Johnstone, 1972). The observations obviously favour a catalytic role for sodium as envisaged by Kimmich (see above).

It is difficult at this stage to make a definitive statement as to which of these two hypotheses is the more acceptable. It is perhaps fair to say that the majority of workers in this particular field of transport studies favour the sodium gradient hypothesis or some modification of that basic model.

Perhaps one should heed Lord Acton's second dictum: 'For every problem there is a solution that is simple, appealing and wrong.'

Further reading and references

Further reading

CHRISTENSEN, H. N. (1970) Linked ion and amino acid transport, in *Membranes and Ion Transport,* Vol. 1, p. 365, ed. E. E. Bittar. Wiley-Interscience, London and New York.

CHRISTENSEN, H. N. (1975) *Biological Transport,* 2nd edn. Benjamin, Reading, Mass.

CLARKSON, D. T. (1974) *Ion Transport and Cell Structure in Plants.* McGraw-Hill, Maidenhead and New York.

HEINZ, E. (ed.) (1972) *Na^+-Linked Transport of Organic Solutes.* Springer Verlag, Berlin.

HOPE, A. B. and WALKER, N. A. (1975) *The Physiology of Giant Algal Cells.* Cambridge University Press, Cambridge, UK.

KIMMICH, G. A. (1972) Sodium-dependent accumulation of sugars by isolated epithelial cells: Evidence for a mechanism not dependent on the sodium gradient, in *Na^+-Linked Transport of Organic Solutes,* ed. E. Heinz. Springer Verlag, Berlin.

MacROBBIE, E. A. C. (1970) The active transport of ions in plant cells, *Quart. Rev. Biophys.,* **3**, 251.

RAVEN, J. A. (1975) Algal cells, in *Ion Transport in Plant Cells and Tissues,* p. 125, eds D. A. Baker and J. L. Hall. North-Holland, Amsterdam and London.

Other references

BRIGGS, G. E., HOPE, A. B. and ROBERTSON, R. N. (1961) *Electrolytes and Plant Cells.* Blackwell, Oxford.

CHRISTENSEN, H. N., RIGGS, T. R. and RAY, N. E. (1952) Concentrative uptake of amino acids by erythrocytes *In vitro, J. Biol. Chem.,* **194**, 41.

CRANE, R. K. (1964) Uphill outflow of sugar from intestinal epithelial cells induced reversal of the Na^+ gradient: its significance for the mechanism of Na^+-dependent active transport, *Biochem. Biophys. Res. Comm.,* **17**, 431.

CRANE, R. K. (1965) Sodium-dependent transport in the intestine and other animal tissues, *Fed. Proc.,* **24**, 1000.

CRANE, R. K., FORSTNER, G. and EICHHOLZ, A. (1965) Studies on the mechanism of the intestinal absorption of sugars. X. An effect of Na^+ concentration on the apparent Michaelis constants for intestinal sugar transport *In vitro, Biochim. Biophys. Acta,* **109**, 467.

CURRAN, P. F., SCHULTZ, S. G., CHEZ, R. A. and FUISZ, R. E. (1967) Kinetic relations of the Na-amino acid interaction at the mucosal border in intestine, *J. gen. Physiol.*, **50**, 1261.

JOHNSTONE, R. M. (1972) Transport of amino acids in Ehrlich ascites cells and mouse pancreas: Evidence against the Na^+ or alkali metal gradient hypothesis, in *Na^+-Linked Transport of Organic Solutes*, p. 116, ed. E Heinz. Springer Verlag, Berlin.

KIMMICH, G. A. (1970) Active sugar accumulation by isolated intestinal epithelial cells: A new model for sodium-dependent metabolite transport, *Biochemistry*, **9**, 3669.

LUCAS, W. J. and SMITH, F. A. (1973) The formation of alkaline and acid regions at the surface of *Chara corallina* cells, *J. exp. Bot.*, **24**, 1.

MacROBBIE, E. A. C. (1962) Ionic relations of *Nitella translucens*, *J. gen. Physiol.*, **45**, 861.

MacROBBIE, E. A. C. (1964) Factors affecting the fluxes of potassium and chloride ions in *Nitella translucens*, *J. gen. Physiol.*, **47**, 859.

RAVEN, J. A. (1967a) Ion transport in *Hydrodictyon africanum*, *J. gen. Physiol.*, **50**, 1607.

RAVEN, J. A. (1967b) Light stimulation of active transport in *Hydrodictyon africanum*, *J. gen. Physiol.*, **50**, 1627.

SCHAFER, J. A., and JACQUEZ, J. A. (1967) Change in Na^+ uptake during amino acid transport, *Biochim, Biophys. Acta*, **135**, 1081.

SCHULTZ, S. G., CURRAN, P. F., CHEZ, R. A. and FUISZ, R. E. (1967) Alanine and sodium fluxes across mucosal border of rabbit ileum, *J. gen. Physiol.*, **50**, 1241.

SCHULTZ, S. G. and ZALUSKY, R. (1964) Ion transport in isolated rabbit ileum. II. The interaction between active sodium and active sugar transport, *J. gen. Physiol.*, **47**, 1043.

SCHULTZ, S. G., and ZALUSKY, R. (1965) Interactions between active sodium transport and active amino acid transport in isolated rabbit ileum, *Nature*, **204**, 292.

SMITH, F. A. (1970) The mechanism of chloride transport of characean cells, *New Phytol.*, **69**, 903.

SPANSWICK, R. M. and WILLIAMS, E. J. (1964) Electrical potentials and Na, K and Cl concentrations in the vacuole and cytoplasm of *Nitella translucens*, *J. exp. Bot.*, **15**, 193.

SPEAR, D. G., BARR, J. K. and BARR, C. E. (1969) Localization of hydrogen ion and chloride fluxes in *Nitella*, *J. gen. Physiol.*, **54**, 397.

VIDAVER, G. A. (1964) Transport of glycine by pigeon red cells, *Biochemistry*, **3**, 662.

WRIGHT, E. M. (1966) The origin of the glucose dependent increase in the potential difference across the tortoise small intestine, *J. Physiol.*, **185**, 486.

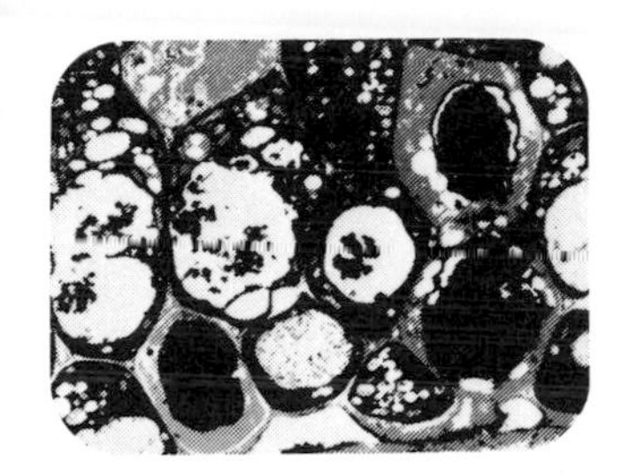

Index

Index